Shoot the Monkey

DO YOU REALLY KNOW WHAT YOU KNOW YOU KNOW?

Gary Barrett

Outskirts Press, Inc.
Denver, Colorado

The opinions expressed in this manuscript are solely the opinions of the author and do not represent the opinions or thoughts of the publisher. The author has represented and warranted full ownership and/or legal right to publish all the materials in this book.

Shoot the Monkey
Do You Really Know What You Know You Know?
All Rights Reserved.
Copyright © 2009 Gary Barrett
V5.0

This book may not be reproduced, transmitted, or stored in whole or in part by any means, including graphic, electronic, or mechanical without the express written consent of the publisher except in the case of brief quotations embodied in critical articles and reviews.

Outskirts Press, Inc.
http://www.outskirtspress.com

ISBN: 978-1-4327-2832-8

Library of Congress Control Number: 2008938455

Outskirts Press and the "OP" logo are trademarks belonging to Outskirts Press, Inc.

PRINTED IN THE UNITED STATES OF AMERICA

Acknowledgements

This book was nothing more than an amorphous idea rattling about in my cranial caverns in the fall of 2006. By the summer of 2008 it took the shape you presently hold in your hands. Whereas the ideas and concepts are solely of my own doing and I must accept credit or condemnation for their presentation, I readily confess that I had significant assistance along the way.

This project could not have evolved into its present form without the invaluable contributions of Britney Spears, Paris Hilton, Michael Moore, Donald Trump, Rosie O'Donnell, "Kramer", Alec Baldwin, Jesse Jackson, Imus, the late D.C. Madam and Anna Nicole Smith, an innumerable cadre of print and broadcast Media Mucky-Mucks, as well as the reigning Political Oligarchy who are faithfully "Doing The People's Business".*

Your influences have forever cemented this work in its proper "time" as well as clarified its "timeless" realities.

I cannot thank you enough.

Gary Barrett

*During the evolution of this project, gasoline soared from about $2.50 to $4.00 a gallon. For this reason some have concluded that our political leaders have "Done Their Business *on* The People" rather than having "Done The People's Business". Not to worry! The Communist Chinese government has come half way around the world to drill the oil in our back yard. I am sure they will give us a good price on this oil now so we can spend our capital developing alternative energy source technologies that they can steal from us later.

Dedication

Before Eliot Spitzer did "Kristen", before Rev. Jeremiah Wright became a loose rail under Barack Obama's Soul Train ride to the White House, before Democrat women realized that sexism still pervaded their party, before Republicans realized they had nominated a candidate that couldn't deliver a speech no matter how well it was written, before Americans on fixed incomes had to choose between heating or eating, and before Britney Spears was celebrated on magazine covers for regaining both her figure and sobriety... there were men of science who loved to shoot monkeys.

This book is dedicated to their relentless pursuit of truth.

Table of Chromosomes

Primordial Soup a.k.a. Introduction

I. Mapping the Helix

1. Pluto and Partial Credit page 5
2. How to Spot Smart People page 15
3. Symbiosis page 23
4. Opposable Thumbs page 35
5. The Pluto Problem page 39
6. Watson's Crick page 43
7. All Politics Are Local page 47
8. Opinions page 59
9. Making a Buck page 65
10. Grading on a Curve page 75
11. tRNA page 83

II. Splitting the Helix

12. The Askin' Why? Chromosome page 87
13. The Teeter Tottered page 95
14. Why? Page 101
15. Becuz' page 107
16. The Continental Divide page 109

17. The Biker Philosopher page 113
18. The Chemical Continuum page 117
19. No Partial Credit page 123
20. The Ultimate Moral Dilemma page 125
21. Presuppositions page 131
22. Wobbly Stars page 133
23. Pablo, the Mayor, and George page 135
24. How to Feel Smart page 141
25. Dark Matter page 151
26. Smart Wars page 161
27. Transcription Error page 167
28. mRNA page 169

III. 'Splainin' the Helix

29. Two Roads page 173
30. Rational Road page 179
31. Lincoln Logs and Erector Sets page 185
32. Global Warming page 189
33. Enigmas of Virtual Morality page 193
34. Designer Drive page 201
35. Creek Jumping page 205
36. Comparative Anatomy page 209
37. Speed Bumps and Pot Holes page 215
38. Tee Shirt Philosophy page 221
39. Moonwalker Conspiracies page 231
40. Final Newsflash page 239

41. Genetic Markers page 245

42. Vestigial Organ page 247

Primordial Soup

I know I am not the brightest bulb in the box.

Yes, I am "one brick shy of a load," "one egg short of a dozen," "half a bubble off," and no, "all my dogs don't bark."

Certainly, better minds than mine are wrestling with the important issues of the day. This allows me to go about my life in obscurity and not have to worry about solving life's weightier problems. I engage in my day-to-day, relatively insignificant endeavors knowing that out there somewhere *intelligent and highly qualified people* are running the world. I trust they are doing a great job with it, because they all tell me their efforts are good, noble, and just.

As they address the issues of fighting terrorism, global warming, feeding the poor, eradicating AIDS, abortion rights, women's rights, restitution for blacks, freedom for the arts, taxing the rich, gay rights, the minimum wage, health care for all, the horrors of hate speech, illegal immigration, affirmative action, free trade, religion in the public sector, gun control, etc., etc., I am filled with a sense of peace that our world is in good hands. After all, we must be in a good place because all of these people tell me they have the answers to these problems.

I watch from the sidelines as important people with fame and power and charisma and education and money and influence manage these responsibilities. I see the *Moral Outrage!* they demonstrate and know they most assuredly walk the moral high ground. Surely things would be so good if the rest of the less intelligent world (people like me) would just follow in their ways. And I, for one, am completely ready to follow them.

Almost.

I just have one haunting question.

I have noticed that ALL of these people are working from the exact same assumption. They ALL assume that their PERCEPTIONS OF MORALITY ARE GROUNDED IN REALITY.

But I don't know how to tell if their assumption is correct.

And although I am not as informed, intelligent, or insightful as them, I think I know enough to guess that if their beginning assumption is incorrect, then their conclusion is probably not too solid either.

So, I set out on my quest for knowledge. I begin as one of the "great unwashed." My first step is simply to ask the question, "How do the intelligent people know for sure that their assumption is correct?"

One last thought.

Not only am I not as intelligent as all these *smart people*, I am not a writer. The idea of putting my thoughts on paper was daunting because of all the great writers and thinkers out there. Take Anna Quindlen from *Newsweek*, for example, who is apparently qualified to be *The Authoritative Moral Arbiter* on any subject she chooses to address. I knew it was a risk competing with the likes of her and her ilk.

But then I became a "Disciple of the Dans."

I learned from Dan Rather that I don't have to be accurate to feel comfortable pushing an idea. I only need to feel justified. If the cause is important enough, Dan can use forged documents about a sitting President's military record to try to scuttle that man's re-election. Who cares if it was the fifth time this issue was being raised and who cares if the President's opponent (who served in Viet Nam) never once had similar military and medical records of his own opened to public scrutiny? As long as Dan felt justified in what he was doing, he did not have to let facts get in the way of what he wrote.

And I learned from Dan Brown in his work *The Da Vinci Code* that people are thrilled when I mix a lot of fiction in with a little truth. Let the lines of reality get blurred a bit. It's okay. It doesn't matter if millions of people reach the wrong conclusions. At the beginning of his book Dan Brown defined for his readers what was accurate (and anything after that was a crap shoot.) In the same spirit, I make this promise: *everywhere I tell the truth it is guaranteed to be accurate.*

Mapping the Helix

Chapter 1

Pluto and Partial Credit

Pluto is dead.

With my apology to Scrooge, this fact must be understood for this story to make sense.

My college transcripts indicate I successfully passed ten hours of physics toward graduation. I have always felt there should be an asterisk next to those credits with a note that says **Partial Credit**. (Much like the Barry Bonds' home run count should be listed in the record book with an asterisk that simply says "Steroids.") I confess. I survived physics solely on the basis of "Partial Credit."

Physics lectures were great because the professor did these over-the-top demonstrations. He taught us "practical" things like the best ways of tearing toilet paper from its roll in order to leave varying amounts of paper remaining. Don't you just hate when there is about a quarter of a roll left and you only need three squares and you yank hard and the whole thing unrolls on the floor? Knowing physics is the answer to that problem.

He opened our minds to understand the "truly amazing," by lying on a bed of nails and letting a big football lineman from the class smash a concrete block placed on his chest. He yanked a tablecloth from under a fully set table and did not break a single plate or glass. (One piece of silverware did fling out among the students, but that only added to the excitement of why physics lectures were fun. My Greek professor never sent tableware at some guy's eye. "Mr. Anderson, please translate pages 122-124 for us and LOOK OUT! HERE COMES A FORK!")

The professor also used humor so effectively to keep our attention that we nicknamed him Prof. Seinfeld. None of us will ever forget his classic demonstration of gravitational pull, when he shot a dart at a falling stuffed monkey and stuck the monkey right in the crotch! A "Standing O" for the professor! Yes, physics lectures were great!

But physics tests were torture. I understood the monkey getting shot in the crotch, but when they asked me to reduce that to numbers and formulas, I was completely lost.

> If h is the height from which the monkey is dropped, s is the speed of the dart, e is the eye, and c is the crotch: calculate the change in funniness, *delta f*, if the monkey is hit in the eye rather than the crotch. Round to the nearest tenth.

I was lost, because I needed a calculator. The only calculator I owned was a slide rule. Who knew the world had changed? In my high school chemistry class a slide rule was *required*, but in my college physics class it was *an object of ridicule*. In a few short years the slide rule was relegated to the status of the abacus, AND NO ONE TOLD

ME! When I pulled out my slide rule for my first physics test it was also the last time I ever used it.

During that test I successfully ignored the snickering of the students around me who were using their high tech battery operated push button calculators with symbols on them that said *smart things* like, "*cos, sin, tan*" (which to this day I don't understand).

What I could not ignore came after the test. I learned that the concrete-block-smashing football lineman had all the maturity and social skills of an eighth grade bully. While the *smart students with calculators* cheered him on, he yanked the slide rule from my briefcase and used it in a most indelicate manner to inflict a wedgie that has left a mark until this day.[1]

After discarding my broken (and now unsanitary) slide rule, I determined to enter the high tech world of calculators. I approached my techie friend, Bob. Bob was an Electrical Engineering major whose great academic challenge was "getting time" on one of the college's few mainframe computers that were reserved for use by *smart people* like Bob who went into electrical engineering. (I never even saw a computer during my college years, but as a Biology major I did get to see dead bodies in my Anatomy and Physiology class.)

So I told Bob my lineman wedgie story. After changing the pants he wet from laughing so hard, he showed me his calculator. It was a magnificent Texas Instruments device that he'd gotten at a great price—only $125! (That's like

---

[1] Though wedgies seem to have withstood "the test of time" with young people, briefcases have also gone the way of the abacus and been replaced by the more casual and much cooler looking backpack.

$41,372 in today's economy.) Anyway, Bob's calculator cost more than twice what my first car cost! So I did what any guy would do. I saved up my money and bought a car.

When the next test came around I had to go from person to person trying to borrow a calculator. This prevented me from being able to actually study for the exam. Eventually I got one from Dan who lived down the hall. "Hi, Dan! I'll give you a ride in my car if you'll let me borrow your calculator for my physics exam tomorrow… Yes, I will buy you a new calculator if that lineman gives me a wedgie with yours."

I picked up the calculator on the way to the test the next day. Because Dan was hung over and running late for his own class he never showed me how to use it. I also had to buy batteries for it because Dan accidentally left it on when he came home from the bar the night before. Dan's irresponsible behavior meant I had to find someone to lend me money for the batteries. All of that made me late for the test.

As I went through the exam, I wore my best *smart person expression*. I pushed buttons on the calculator and then wrote something on the test. I would furrow my brow and act like I was working really hard. I tried to maintain a *smart person rhythm* between writing and pushing buttons. Write, write, write…push, push, push. I made sure the football guy saw I had a calculator, but I never made eye contact with him in case it gave him any new ideas about wedgies (or suppositories!). Not having a clue about the test, I quickly became enamored with Dan's calculator: Write, write, write…push, push, push… "Hey, if I punch .07734 and turn it upside down it looks like 'hello'. I wonder if I could ask that cute girl next to me for a date with this thing. How could I make it say…are you busy tonight?" When the cute girl got up to turn in her exam, I remembered I was supposed to be taking a test, so I wrote some more. After the bell rang I brought my exam to the front with my best imitation *smart person swagger*. And upon leaving the room I sprinted a mile just in case that football guy was behind me.

At the next lecture they returned our exams. I fully expected to fail, because I knew I was clueless on the test. Imagine my surprise when I received a "C". How do you pass an exam in which you answer no question correctly? Simple. Partial Credit. Those *smart people* in the Physics Department had kindly agreed to give us *dumb people* (i.e. Biology Majors) credit for whatever part of the problem they decreed we got right.

Back then I actually believed that somewhere on campus there was a *smart person grader* who generously labored over our *dumb people* exams to determine just how much of each problem we actually got right. Of course, such kindness from the *smart person grader* was well received by me, because I needed those physics credits so the *smart people administrators* would give me my diploma.

But now I know better.

That partial credit "C" that I got on every test from the *smart person grader* was a bribe to keep me quiet. He didn't know the answers either! If I would stay quiet about the fact that I learned nothing in the class (other than the knowledge that shooting a falling monkey in the crotch was a hilarious sight gag) they would pass me with a "C". That way, I got my diploma, and they got to keep their *smart people jobs* so they could get paid for shooting darts at monkeys.[2]

How do I know now they didn't know the answers either? Two things. First, if they were really that intelligent, they would have saved the $125 they spent on calculators and invested it in Microsoft. By now they would be gazillionaires!

Second, Pluto is dead.

Newsweek said so. The September 4, 2006 issue arrived in my mailbox with a cover story on the death of Pluto. I didn't think it was possible. But then, Ray "I-lost-the-keys-to-a-thousand-school-buses" Nagin was also on the cover. Since I didn't think he could get re-elected as the mayor of New Orleans, I concluded anything was possible.

With trepidation I read "Requiem for a Planet." It was there I realized I had been taken for a ride all these years. The article had the requisite pictures and diagrams of the planets. It also had pictures of *SMART PEOPLE ASTROPHYSICSTS!* One picture showed Clyde Tombaugh, Pluto's discoverer, looking very intelligent while standing next to a telescope, peering up at the night sky. He had a pen and pad in his hands like he was going to do some *smart writing*. But this guy isn't so intelligent. Even I know YOU HAVE TO LOOK THROUGH THE TELESCOPE in order to see anything in it. Imagine buying a new telescope with instructions that say, "Point telescope at sky and stand next to it looking up with an intelligent expression. Having pen and paper in your hands is optional, but *it will make you look smarter.*"

---

[2] Google "Shoot the Monkey" and you will find they are still using this ploy 'til this day.

Not only was Tombaugh not looking through the telescope, his pad of paper was only about the size of a deck of cards. Everyone knows s*mart people* do LONG EQUATIONS. He didn't even have enough room to get partial credit on that pad.

In another picture, also designed to make us aware that WE ARE NOT SMART, was a room full of *smart people* deciding the fate of Pluto based on its size, shape, and lack of orbital dominance.[3] Everyone was holding up voting cards. The caption says, "It's history. Astronomers vote in Prague". [#3] The scene is reminiscent of the 2000 presidential election and south Florida county officials holding up the voting cards with the hanging chads. If these astrophysicists were really intelligent, they wouldn't have been using these cards because *smart people* know that the results from such voting cards are not accurate. These cards produce too many disenfranchised voters, and no matter how many times you recount the vote, even if you use your fancy push button calculators, you still can't net the results you're trolling for.

In a related article in *Readers Digest*, someone commented that maybe Pluto would not have lost its status as a planet if it had a more serious name. Oh, really! Then how do you explain that 'Uranus with its 27 moons' is still taken seriously?

But the real give-away that I had been taken for a ride all those years was in the pull quotes, not in the pictures. One said, "The sad truth is, Pluto's claim to planethood has been shaky ever since its discovery." *

---

[3] I checked the picture carefully and you and I are not in it. After all, if you were smart enough to be in that picture, would you be reading this?

Hold the phone!

Back the truck up!

Run that by me again!

"The sad truth is, Pluto's claim to planethood has been shaky ever since its discovery."

You know what that means? THEY NEVER KNEW FOR SURE IF PLUTO WAS A PLANET OR NOT! But you and I, who are not part of the *smart people,* were supposed to believe what they said anyway. And don't ask questions!

I can just imagine this conversation with my grade school teacher Mrs. Miller.

Teacher: Class, who can tell us how many planets are in our solar system?

Peggy: Nine

Teacher: Very good, Peggy. Now who can tell us where our planet Earth fits in?

Donna: Earth is the third planet from the sun after Mercury and Venus.

Teacher: Very good, Donna. And who can tell us…

Me (interrupting): Excuse me, Mrs. Miller. I really do not think there are nine planets. Based upon its size, shape, and lack of orbital dominance, I do not think Pluto should be considered a planet.

Teacher: That's very nice, Gary, but there are *smart people* called astrophysicists and they have push button calculators. They told us Pluto is a planet and if they say it, then it must be true. These *smart people* write our science books. Our book says, "Pluto is the ninth planet." Are you smarter than the *smart people?*

Me: Oh no, Mrs. Miller. I only have a slide rule, so they must be right.

And so, dutifully, me and my entire Baby Boomer generation were tested and graded upon what the *smart people* told us was true even though THE *SMART PEOPLE* DIDN'T KNOW FOR SURE THEMSELVES WHAT THE TRUTH WAS!

Well, the cat is out of the bag, and I am not gettin' suckered in again by the *smart people*! The second pull quote in the *Newsweek* article said, "Researchers say they have found the first proof of the existence of 'dark matter'." [+] Yeah, right! Black holes! So dense that their gravitational pull doesn't even allow light to escape! You'll never see one, but you can trust the *smart people* that they really do exist.

Not this time, buddy! I believed you on Pluto, but I am not buying Black Holes. "Fool me once, shame on you. Fool me twice, shame on me." Although I am wondering…if that monkey was falling into a black hole, would the dart still hit him in the crotch?

The vote on Pluto's status as a planet was nearly unanimous.

Chapter 2

How to Spot Smart People

Right next to my "Pluto Died" issue of *Newsweek* I have a copy of the October 9, 2006 *TIME*. The cover photo features side-by-side, half-faced shots of an extra from the original *Planet of the Apes* and a baby picture of Uncle Fester from *The Addams Family* TV show. The cover story is titled, "How We Became Human." Actually, the cover story should be called, "We're Smart, Too!"

Buried in the article is this statement: "Although the news was largely overshadowed by the impact of Hurricane Katrina, *which hit the same week*, the publication of a rough draft of the chimp genome in the journal *Nature* ..."[#] (emphasis mine). You know what that means? Their *TIME*-ly article was ALREADY 14 MONTHS OLD!

So why were they printing old news? The timing of the story gives the answer. A month earlier, *Newsweek* ran its Pluto issue. That issue had special graphics and pictures and quotes of *smart people*. And because nobody wants to be dumb, *TIME* needed to run its own issue with special graphics and pictures and quotes of *smart people*.

It went like this at the boardroom of *TIME* on September 5:

Big Important Boardroom Person: Okay people, *Newsweek* kicked our butts this week with that Pluto thing. We need a cover story so that we can be smart, too. What have you got?

Little Un-important Boardroom Person: Here's a monkey story that got lost in The Hurricane.

Big Important Boardroom Person: Can we do special graphics with pictures and quotes of *smart people*?

Little Un-important Boardroom Person: We have a graphic of monkeys in a tree and a picture of some guy holding a skull.

Big Important Boardroom Person: That will do. Put it on the cover next month.

Never having been accused of being one of the *smart people* myself, it was hard, at first, to spot someone who was really intelligent. Every kid from the playground knows how this works.

Terry: You're a stinky ugly jerk!

Billy: Takes one to know one!

Actually, I learned how to spot *smart people* by being cheap. Now THIS is something I HAVE been accused of many times!

My education began in that great icon of American capitalism, the convenience store gas station. These are architectural marvels, because they can house an ENTIRE SHOPPING MALL IN A SINGLE GARAGE! They have food courts, auto parts, drug stores, hardware, clothing, bedding, gifts, electronics, literature, sporting goods, novelties, artwork, alcohol, tobacco, gambling (who needs Vegas?), and home entertainment products. Where else will you find a sign like the one near my home that simply says, "HOT FOOD LIVE BAIT"? How many restaurants would ever pass inspection if hundreds of leeches were just a few feet from the food? Imagine a *smart person* in that restaurant.

Smart Person: That "George Washington" special sounds interesting. What's in it?

Waiter: Leeches. Would you like the house wine with that?

One of the joys for a cheap person in a convenience store is the price available on used videos. Because VHS is heading the way of the slide rule, movies that used to cost $39.99 can now be purchased for $3. More importantly, *smart people* videos can still be found!

On a recent outing, I found a movie nominated for 5 Academy Awards and the winner of 2 Golden Globes. The video jacket said, "In this intoxicating, *intelligent comedy*...the comically mismatched pair soon find themselves drowning in wine, women... and laughter." (Italics mine.)

More importantly, the *smart people* review from *Newsweek*

said, "WONDERFUL, HILARIOUS...DELICIOUSLY BITTERSWEET." Not to be outdone, the *other smart people* review from *TIME* said, "BY FAR THE YEAR'S BEST AMERICAN MOVIE." *

Feeling confident that I was on to something good, because this was an INTELLIGENT COMEDY and all the *smart movie reviewers from the smart people magazines* said it was great, I was feeling pretty smart myself when I paid my $3 and got out of there with the movie before someone truly intelligent bought it.

That night I sat down to watch the movie, waiting to split a gut. I didn't dare drink milk during the movie because I knew AT ANY MINUTE I would be laughing so hard it might come out my nose. I try and avoid this experience at all costs because once in the 7th grade cafeteria the class bully slipped on a butter patty while I was drinking milk and I laughed so hard milk came out my nose. It really hurt! Of course, it hurt even more when the bully poured some milk back up my nose.

About half way through the movie I realized I hadn't laughed yet. So I waited longer, figuring I probably would be turned to hysterical mush by some climactic ending.

I am still waiting...

NOTHING!

No chortle, chuckle, snort, guffaw, snicker, or smile. I was feeling like I had been cheated out of my $3, when it hit me. This is INTELLIGENT COMEDY! I thought when I paid my $3 that I was purchasing a night of laughter. But when you buy INTELLIGENT COMEDY laughs are not included.[4] It was then that I understood how to spot *smart people*. *Smart people* know how to get us to happily give them money in return for nothing!

Once I knew how to spot *smart people* I started seeing them everywhere.

First, I spotted the *smart people* at the **International Star Registry**. To show how much you really love someone, you can name a star after them! For only $379 you will get A PIECE OF PAPER that claims some star on some chart has your loved one's name on it. You of course will never actually get the star. And if you want to actually see your star they will sell you a telescope for an additional $697. My friend Butch got the telescope. It came with these directions. "Point telescope at sky and stand next to it looking up with an intelligent expression. Having pen and paper in your hands is optional but *it will make you look smarter.*" A week after you die, your kids will read in *Newsweek* that they renamed your star. The pull quote will say, "The sad truth is, calling this star Jennifer Morganstein has been shaky ever since its discovery."

Then I spotted the *smart people* of **Multilevel Marketing**. To make a fortune and retire young it only costs $5,000 to get in and START YOUR OWN BUSINESS! This seems like a steal because everyone knows that it costs about a billion dollars to start a McDonald's or Starbuck's. (If I could *afford* to start one of those, I wouldn't *need* to start one of those.) And what do you get for your $5,000? You get the right to say, "I own my own business." Of course,

---

[4] If you want laughs, buy SpongeBob Squarepants. I recommend the Sea Stories DVD with multiple endings for the Flying Dutchman. Drive about twelve hours to a friend's house so you are really tired. Watch the cartoon about midnight and just see if you can keep from splitting a gut. And DON'T drink milk while watching.

you never actually make any money until you get 62,000 other people to cough up the dough to say, "I own my own business."

And then I spotted the *smartest person in the world!* His name is **Sam**. Sam is so smart that he charges people to shop at his store. Every other store in town PAYS FOR ADVERTISING to entice you to shop at their nicely decorated stores, but Sam CHARGES YOU to shop in his ugly warehouse! And what do you get for the "Membership" fee of $148? You receive your own SPECIAL CARD that lets you check out. Let's see, 372,539,402 "Members" paying $148 each means Sam made over $849 trillion without selling one item. And next year, they are all going to willingly renew their "club dues"! Clearly, Sam must be the *smartest person in the world ever!*

In reality, the *smart people articles* in news magazines that entice us to happily pay for nothing are highly overrated. Nobody reads them anyway. Who wants to slog through stuff like this from the *TIME* article?

> A discovery published online by *Nature* last month suggests Neanderthals may have made their last stand in Gibraltar, on the southern tip of the Iberian Peninsula, surviving until about 28,000 years ago— and possibly even longer. [+]

SNOOZER!

We are Americans and we demand to be entertained, not educated! How else can you explain that in a recent poll, it was determined that the average American knows all of the Simpsons, who was last voted off of American Idol, why Jessica and Nick broke up, that Britney doesn't always know how to operate a car seat, that Jennifer still loves

Brad, that Tom went crazy on Oprah's couch, and that Mel says stupid things when he is drunk, but that same American CAN'T NAME TWO RIGHTS LISTED IN THE BILL OF RIGHTS!

This is why educational articles about monkeys in *TIME* magazine take up only 2 1/2 pages (including special graphics and pictures and quotes of *smart people*) but full-length entertainment movies about monkeys taking over the world have17 sequels.

Chapter 3

Symbiosis

Through NATURAL SELECTION driven by RANDOM MUTATION there has EVOLVED over many, many years a SYMBIOTIC RELATIONSHIP between *smart people writers* and *smart people scientists*.

Symbiosis is a mutually beneficial living arrangement between two different organisms. The scientists provide the writers with something to write about so that the *writers get our money* through the sale of their magazines. The writers put the scientist's pictures in their magazines so the *scientists get our money* through big government research grants. This is *our tax money* doled out in pork barrel projects (a.k.a. "earmarks") designed by lawmakers to "bring home the bacon" to their own state for the purposes of re-election. Previously, this practice was isolated within the Democratic gene pool but through genetic migration has recently become well implanted in the DNA of the Republicans. Now when it comes to spending our money no one can tell the difference between the two.

The symbiotic relationship between writers and scientists is carried out with a straight face while they openly admit that they don't know as much as we think they know. If you and I did something like this they would say we were in cahoots with one another and people were being ripped off. These people get away with it because they are intelligent, and as we have learned, *smart* people know how to get us to happily give them money in return for nothing.

This symbiosis (literal meaning "in cahoots") is maintained through intricate and delicately balanced processes known as **Statistical Overload** and **Linguistic Charades**.

Statistical Overload is a standard journalistic ploy whereby the writer utilizes SO MANY NUMBERS that he *sounds smart* but says nothing. The goal is to get the person to quit reading the article, believing that *they are not smart enough* to understand it, before they realize that the writer really has nothing to say.

There are three steps to this method.

Step One: Use a **picture and a clever title** to get the reader's attention.

Step Two: Throw in a **bunch of numbers** that so confuse the reader he decides to simply "skim" the article because he feels he is not smart enough to "get it" anyway.

Step Three: Throw in a **second bunch of numbers** so the reader feels so dumb he quits skimming and starts looking for a MORE ENTERTAINING part of the magazine. (The trick here is that because Step Two sent the reader into "skim" mode, you need a real lot of numbers so he or she will see them even while skimming.)

The Monkey Story in *TIME* is a perfect example of this technique.

Step One: (pages 44-45)

> Reader *sees*: An interesting composite picture of the extra from the *Planet of the Apes* and Uncle Fester as young adults.[5]
>
> Reader *reads*: "WHAT MAKES US DIFFERENT?"
>
> Reader *thinks*: "This looks interesting."

Step Two: (pages 46-47)

> Reader *sees*: A graphic of the "evolutionary tree".
>
> Reader *reads*: 3 billion NUMBER OF BASE PAIRS
> 1.23% PERCENT THAT ARE DIFFERENT IN THE CHIMP GENOME
>
> Reader *thinks*: Huh? I better just skim this.

Step Three: (pages 48-49)

> Reader *sees*: A picture of three scary skulls.
>
> Reader *reads*: "It was estimated that humans had 100,000 genes. When we got the genome, the estimate

---

[5] In this picture Uncle Fester is sporting a trendy buzz cut as opposed to his better known bald guy look.

dropped to 25,000. Now we know the overall number is about 22,000, and it might even come down to 19,000."

Reader *thinks*: I wonder if Britney learned how to use a car seat yet...

The point of the Statistical Overload technique is to PREVENT THE READER from ever finishing the article. That is because if we ever do finish it, we will discover the article is a bunch of Hooey.

Just take a look at these jewels on page 50:

> Scientists have long used the average difference between genomes as a sort of evolutionary clock because more closely related species have had less time to evolve in different directions.

Then, after a TITILLATING DISCUSSION about the cohabiting of various species (even in the scientific world, "sex sells"), this pearl is slipped in:

> All of that depends on the accuracy of fossil dating and the reliability of using genetic variation as a clock. *Both methods currently carry big margins of error.* (Emphasis mine)

Now that is a sure fire admission that all along they have been SAYING NOTHING! In the political world it would be reported this way: "In our latest CNN poll Barack Obama is leading John McCain 53-47% with a margin of error of plus or minus 168%."

Further along on this page is a discussion about the differences between Neanderthals and us. Although they were intelligent, they were not as intelligent as we are. That is why they are called Neanderthals, which literally means "Slide Rule Man." The page has a picture of a *smart person molecular biologist* holding a skull. The article is called "What Makes Us Different?" and here we read that this *smart person* "...admits that he still hasn't learned much about what distinguishes us from our closest cousins." HO, HO, HOLD ON THERE! You know what that means? The *smart people molecular biologists* are no more sure about our differences from Neanderthals than the *smart people astrophysicists* were sure that Pluto was a planet!

Whereas the journalist uses Statistical Overload to throw us off track in the articles they write, the *smart people scientists* use Linguistic Charades to totally fog up their lack of any real knowledge.

Linguistic Charades is the classically simple but completely effective technique employed by scientists to conceal the fact that they know less than they are letting on. This is an excellent communication skill which ALLOWS SOMEONE TO SEEM SMART WITHOUT HAVING TO KNOW ANYTHING. This approach involves the use of one key word or phrase. By using the EXACT word or phrase, they immediately *sound smart*. The beauty of this skill is that it is not limited to the sciences and so allows ANYONE to sound smart about ANYTHING if they simply use the exact word or phrase.

For example, in sports, the word is *momentum*. You can use the word in many different ways. In fact, for the more skilled person playing Linguistic Charades, it becomes a personal challenge to see how many different ways they can utilize the same the word.

Interrogative Sentence: Do you think the Huskies have the *momentum* to go all the way this year?

Implied Question: I wonder if the Cardinals' *momentum* will carry them through the playoffs.

Declarative Statement: I think the *momentum* favors the Raiders at this point.

Completely Generic Question: Who do you think has the *momentum*? (The advantage to this question is you do not need to know the sport or even the team names to sound like you know what you are talking about.)

The reason the word *momentum* works so well in sports is that all sports lovers are armchair quarterbacks. They have an insatiable desire to show how *sports smart* they are. So they will gladly take your cue and give you their opinion. After that, you only need to NOD AND LOOK INTERESTED and they will think you are smart too, even though *you know you are as clueless as a biology student taking a physics test.*

In politics, Linguistic Charades have a rather short half life, because the people and topics change so rapidly. The correct phrase for '06 was *culture of corruption*. During the Viet Nam war it was *quagmire*. Perhaps the best known Linguistic Charade (and my all time favorite) was *gravitas*. This word had a lifespan of exactly one presidential campaign. The word was never used before the 2000 campaign nor have we heard it since. That is because before the votes were even counted it was replaced with "*stole the election.*"[6]

With *gravitas*, the sentence structure was not as important as the delivery. The most effective approach was the slightly delayed use of the word with a pained expression on the face, as if the person was searching their vast cranial lexicon for the EXACT right word. "The big question that must be answered is, does George Bush have the…the…ah…*gravitas* to serve as President?"

On talking head shows with more than one talking head, it is usually a race to see WHO CAN INTERJECT THE LINGUISTIC CHARADE FIRST during the show. After that, they can simply "nod and look interested" and not have to say anything more for the remainder of the show. That way they are not at risk for saying something totally dumb that will get replayed endlessly as "controversial" on every news show for the next three weeks (a la John Kerry's classically stupid line "stuck in Iraq" which he then had to try to pass off as a bungled joke).[7]

During the 2000 campaign, on Chris Matthew's *Hardball*, Judy Woodruff set a LINGUISTIC CHARADES WORLD RECORD THAT STILL STANDS. Her time was an amazing 1.6 seconds!

Chris Matthews (opening his show): I'm Chris Matthews,

---

[6] This was the election where Al Gore found some spare time between inventing the Internet and saving the world from Global Warming in order to take a run at being President.

[7] Because John "I fought in Viet Nam" Kerry was the *Intelligent Candidate* who understood the "nuances" of every issue, it was very difficult for him to convince the rest of us dummies that this was a bungled joke. Intelligent humor doesn't make us laugh so we can't tell the difference between a bomb and a real gut buster.

let's play Hard...

Judy Woodruff (interrupting and pounding table): Bush! Gravitas!

Chris Matthews: Wow, Judy! I think you just set a LINGUISTIC CHARADES WORLD RECORD! And you folks at home saw it here on *Hardball*.

For scientists, the classic LINGUISTIC CHARADE is *dark matter*. Whatever scientists don't know is always located in the *dark matter*.

Notice the full text of *Newsweek's* "Dead Pluto article" that produced the aforementioned pull quote. It states:

> Researchers at the University of Arizona said they have found the first proof of the existence of *"dark matter"*— an invisible substance, unlike any known atoms or particles, whose gravity holds galaxies together. [#] (Emphasis mine)

Translated into common English, this says, "We are completely ignorant as to why the galaxies haven't blown to bits about a billion years ago."

Now compare that quote with these excerpts from the *TIME* monkey story. After describing how mapping about 47 million genes still didn't explain the differences between monkeys and humans, they concluded it was not the genes that made the differences but the "molecular switches" that control these genes. Notice what they say about these switches:

> Those molecular switches lie in the noncoding

regions of the genome—once known dismissively as junk DNA but lately rechristened the *dark matter* of the genome…Coding regions are much easier for us to study…But it may be the *dark matter* that governs a lot of what we see.* (Emphasis mine)

Translated into common English, it says, "After 137 years and 82 billion tax dollars, we are still in the *dark* about what makes us different from monkeys because we have been looking in the wrong place all this time."

Science Comes Home...

Chapter 4
Opposable Thumbs

America's Funniest Videos is NOT intelligent comedy! We know this for two reasons. First, because 93% of their videos feature someone whose pants fall off. Only a moron could find the 632nd time someone's pants fall off as still funny.[8] Second, *AFV* is not intelligent comedy because the AUDIENCE LAUGHS, and we know intelligent comedy does not generate laughter.

On October 30, 2006, the host introduced a segment on animal clips by reminding the audience that OPPOSABLE THUMBS distinguish us from animals.

This is important to note because the dumb people didn't think this up. They were too busy making videos of people's pants falling off. They learned it from the *smart people* in their grade school science books and articles like the *TIME* monkey story which states,

---

[8] Even shooting the monkey in the crotch was getting old after 546 times.

> As paleontologists have accumulated more and more fossils, they have compiled data on a long list of anatomical features, including body shape...and *opposable thumbs*.[#] (Emphasis mine, but you knew that.)

By now, everyone has been taught that opposable thumbs set us apart from other animals. In fact, there are reported instances of Higher Functioning Fifth Graders using *opposable thumbs* as their own means of playing LINGUISTIC CHARADES in the classroom. This works especially well with insecure first-year teachers.

First-Year Teacher: You can see from this evolutionary tree that the *smart people* made for us, that we are more like a monkey than a snail. That is because snails have only one foot and monkeys have two feet like us.

Fifth Grade Students: Oooooh.

First-Year Teacher: Now, if we ever found snails that were conjoined twins, then they would have two feet and be just like us. But because the fossil record doesn't show conjoined snails, this proves we came from monkeys.

Higher Functioning Fifth Grade Student: If snails had *opposable thumbs*, would they have more *gravitas* than George Bush?

(Now the truth is, even without opposable thumbs, snails have more *gravitas* than George Bush. But because this teacher has been trained to pay her union dues to the NEA so they can tell her how to vote, she has never had to think for herself politically so she has no idea what the student has just said. Her response is to switch the subject.)

First-Year Teacher: Well, that is all the time we have for science class today. I want every one to take out their Spelling Books.

Higher Functioning Fifth Grade Student: Teacher, I was wondering, how do you spell the plural form of "potato?"

If the *smart people* can't agree on Pluto, and the *smart people* can't define what makes us different from monkeys, then the *smart people just might be wrong about opposable thumbs.* I realize that challenging the conventional wisdom of the ages handed down to us from the *smart people* makes me look even more foolish than Chicagoans waiting for the Cubs to win the World Series. But I'll take that risk because I know something the *smart people* don't know. I KNOW SOMEONE BORN WITHOUT THUMBS! (Well, actually we lost track of one another after high school, but he really did have no thumbs.)

If having opposable thumbs is what separates us from the animals...then this guy in my school was technically not human. He has to be considered a dramatic throwback on the evolutionary tree. But he didn't look or act like he came from the Planet of the Apes. He wasn't the Wolfman. And he certainly didn't grow a shell like a snail. In fact, he was completely human. He just happened to be a human without thumbs.

Which leads us to the simple revelation that even a partial credit, slide rule using, likes-to-laugh-at-funny-movies dumb person like me could maybe guess that WHAT MAKES US DIFFERENT FROM OTHER ANIMALS IS NOT OPPOSABLE THUMBS!

****Newsflash!****

While writing this chapter, my November 6, 2006 issue of *Newsweek* came across my desk. Believe it or not, the article on pages 60-61 is called "Plotting Pluto's Comeback." Two things must be noted.

The pull quote states: "Some astronomers want to reclaim the status of planet for the distant ball of rock and ice." Do you know what that means? *The smart people still don't know what to do with Pluto!*

When dealing with the motives for the change in the first place, the article states: "…to put it bluntly, funding is involved…" Hmmm…this whole thing sounds suspiciously like what was said in the first paragraph under SYMBIOSIS.

Chapter 5
The Pluto Problem

Smart people writers are trained to understand their readers. They know that while most of their readers will fall prey to Statistical Overload, there are always a few readers *pretending to be smart* who read the last paragraph of their "smart" cover article before going to the article about Britney's car seat problems. That way they can impress regular dumb people at the office who jumped directly to the Britney story.

Pretending-To-Be-Smart Dumb Person: Did you read *TIME's* cover story on, "What Makes Us Human?"

Regular Dumb Person: Nah, I couldn't get through all those numbers. But I did read about Britney.

Pretending-To-Be-Smart Dumb Person: They concluded that in a few years they will know everything.

Regular Dumb Person: You know, if someone could design a car seat easy enough for Britney to use, they would make a fortune.

This type of reader is known to the *smart people writers* as a POMPOUS LAST PARAGRAPH PRETENDER. Because of them, the *smart people writers* know that their last paragraph must appear to reach some significant conclusion. That is why the *TIME* monkey story ends with this paragraph:

> For most of us, though, *it's the grand question about what it is that makes us human* that renders comparative genomes studies so compelling. ...yada yada ...random process ...mumble mumble ...genetic changes ...etc. etc. 3.5 billion years ...Within a few short years, we may finally understand precisely when and how that happened. [#] (Emphasis courtesy of Yours Truly)

And to this grand conclusion we can only say, "LET'S GIVE PARTIAL CREDIT, WHERE PARTIAL CREDIT IS DUE."

These *smart people scientists* really have something to contribute. I have a *CSI* kind of guy who hunts near my home every year. I'll call him Terry. (Everybody else does so it works for me.) Terry is employed by the government as a blood splatter specialist. His position probably has a cool name like HEMOSPLATOLOGIST, but I don't know what it is. Recently he told me he was being called back to court to deal with a cold case from 30 years ago. I asked him how technology has impacted his work over the course of 30 years. DNA testing was one of the major advancements. Thirty years ago they only had A-B-O testing. This gave very low percentages on anything definitive. It could tell you who did not do the crime, but it couldn't point the finger at who did do it. Now, with DNA testing, they are able to identify miscreants with virtually absolute certainty. (Think "blue dress"... Don't think "O.J.")

Medical advancements are also anticipated as genomes are deciphered. Malaria, AIDS, viral hepatitis, and Parkinson's are but a few conditions being named as potential areas of breakthrough. For these very real accomplishments we thank the *smart people scientists*.

But for all of their *smart people knowledge*, there is one problem. I know it will cause some people great stress to consider this, but it is true. The *smart people scientists* are NOT gods. Tackling the *"the grand question about what it is that makes us human"* they will learn they can't answer that question solely at a chemical level.

The first problem they face will be **The Pluto Problem**.

Eventually, they will map the genomes down to the order of specific molecules. They will be able to locate billions of molecules in the DNA strands. And then they are going to tell us what makes us human. YEAH, RIGHT! If our solar system—made of nine objects revolving around the sun—stumps the *smart people* as to what defines a planet, do we really think they are going to be able to define something made of billions of molecules? I DON'T THINK SO!

They will be able to say how we differ from monkeys, but being different from monkeys doesn't make us human, it just makes us different from monkeys.

Chapter 6

Watson's Crick

Centuries ago some *smart person philosopher* sought to prove his own existence with the concept, "I think, therefore, I am." I don't know how smart you have to be to question your very existence, but I proved my own on a number of occasions. My proof was this, "I hit my thumb with a hammer, therefore, I screamed."

Whenever I hit my thumb, I have no question that I exist. SOMETHING got smashed. SOMETHING felt pain. SOMETHING screamed. And SOMETHING heard the screaming. All of those SOMETHINGS were me. They prove that I exist and that I am made of matter. The *smart people scientists* will soon be able to map my blueprint called DNA that uniquely arranges my SOMETHING to make me, ME.

But when they think that they can define my humanity by mapping my molecules, they are paddling their canoe up Watson's Crick. That crick leads to a dead end, and there ain't a paddle big enough to go any further.

Consider these definitions in some future Biology book:

> Monkey: n. a tool-using primate whose DNA pattern causes him to say, "OOH OOH! EEH EEH! AAH AAH!" when he hits his thumb with a hammer.
>
> Human: n. a tool-using primate whose DNA pattern causes him to say, "OH %@#$*&+!" when he hits his thumb with a hammer.

My humanity is more than a mapping of molecules, just as Texas is more than a page in a Rand McNally atlas. Mapping my molecules totally overlooks the subtle humor in my term "Watson's Crick." It misses the joy I bring to all those around me. It cannot encompass the deeper purpose being pursued in my life when I am watching the NFL while eating M&M's. In short, my humanity is not material. When the *smart people scientists* have totally mapped my matter they will have accomplished an amazing and useful goal. But they won't have touched my humanity.

I learned this reality through America's free public education system. Being an Average American I study at this school for about 22 hours a day. It's called the University of Television.

In the course, *Touchy Feely Thoughts About Large Impersonal Conglomerates 101*, the one-minute class was being taught by a visiting lecturer from the DOW Chemical Company. With effective use of visual aids, the lecturer informed us that the *smart people scientists* at their company have discovered a NEW ELEMENT! It is called the Human element. It is designated "Hu" on the Periodic Table of Elements. And when this element is injected into the equation, everything changes!

When the *smart people scientists* who are paddling up Watson's Crick hit the dead end and still haven't found the Human element (Hu) buried somewhere in our DNA, we can only guess where they will tell us it's located.

> Today scientists announce they have paddled their way to the end of Watson's Crick. They finally mapped every molecule in the human body. As of yet they still can't explain what makes us human. They speculate the answer to that question lies in something called *dark matter*.
>
> Even though they never really came up with the answer, the *smart people graders* have given them a "C" on this project thanks to the wonders of Partial Credit.

The next goal of the chemists at DOW chemical is to find the "Monkey Element" (Mk). It is believed that when this element is injected into the equation, everything goes bananas.

****Newsflash!****

While writing the previous chapter (November 9, 2006), Shepard Smith reported that Britney Spears' soon-to-be-ex-husband K-Fed's Lawyer's Publicist (yes, his lawyer has a publicist!) said they were not sure yet if he would sue for custody of their two kids on the grounds that Britney was an unfit mother. Their potential plea would be that she did not know how to use a car seat.

I am NOT making this up!

This news is going to make the *smart people writer's* work a whole lot easier. The dumb people will be so anxious to read about how Britney's car seat problems are going to impact her divorce they will only need a few statistics to skip to the more entertaining reading. Statistical Overload will be a breeze.

Dumb person reads: Scientists believe that the eight remaining planets...

Dumb person thinks: Skip the planets! I wonder if Britney will have to go to car seat training to keep her kids...

Chapter 7
All Politics Are Local

Years ago, a *smart person politician* tipped us off to the idea, that "All politics are local." Until recently, this has been an accepted fact of American lore, in the grand traditions of "Cleanliness is next to godliness," "Moss grows on the north side of trees," and "All politicians are liars." But the mid-term elections of 2006 proved this understanding of politics is wrong. The bit about politicians being liars is accurate, but the idea that all politics are local was summarily dismissed.

More conclusive than anything seen on *Mythbusters*, the election proved that the voters were disgusted (rightfully so) with the performance of the Republican Party as a whole. One exit poll indicated that 74% of the voters were concerned about corruption, 68% were concerned about the war in Iraq, 72% were concerned about illegal immigration, and 59% felt that we weren't investing enough tax dollars to study the environmental impact of bovine flatulence on global warming. These were decidedly not local issues.

(For those suffering Statistical Overload about now, the next news about Britney is found at the end of this chapter.)

Even people who are only "slide rule smart" know that politics are not confined to the government. There are office politics, boardroom politics, family politics, church politics, neighborhood politics, dating politics, teenage politics and so on. Politics are as easy to find as salsa in a Mexican restaurant.[9] Because politics are universal to the human experience, perhaps hidden somewhere in the *dark matter* of our political DNA is a hint of what makes us truly human. Once the *smart people scientists* have studied our POLITICAL DNA they will discover that all politics are not local but ALL POLITICS ARE MORAL.

Regardless of where you find it, the heartbeat of all truly passionate political discourse is *Moral Outrage!* This is the great rising of emotion that accompanies the perception of injustice, unfairness, indecency, or unpunished wrongdoing. Without *Moral Outrage!* political wrangling is BORING.

Consider how *Moral Outrage!* makes family politics so much more interesting:

Father: Lisa, all of your brothers and sisters had a curfew of 10:30 on their first date. It is our family rule.

Lisa: Yes, Father. I will be home at 10:30.

BOOOORRRING!

"Kick it up a notch" with a little *Moral Outrage!* and

---

[9] Doesn't it even say in the Bible somewhere, "Wherever two or three are gathered, there will be politics also"?

"BAM!" family life gets interesting:

Father: Lisa, all of your brothers and sisters had a curfew of 10:30 on their first date. It is our family rule.

Lisa (screaming, whining, crying, stomping and feeling a deep sense of Moral Outrage!): 10:30? All my friends get to stay out 'til midnight! I AM ALMOST 14! I should be allowed to decide these things for myself. You don't like Jimmy just because he sells drugs, but you smoked pot when you were a kid! You are a hypocrite! You don't trust me! All the other kids will laugh at me! It's not fair!

At this point, the father has one of two choices. He can stick by the family rule and have a moody adolescent girl sneering at him for the next two months or give in to her demands and have the older kids who all had 10:30 curfews sneering at him for the next two months. So he does what any normal father would do. He tells her to talk to her mother.

In the highly respected, scientific, and peer reviewed *Readers Digest*, the article entitled "Animal Einsteins" (November 2006) says:

> ...researchers have found animals communicating complex ideas, solving problems, using tools and expressing their feelings - behaviors once thought to be uniquely human.

Chalk this one up as another OOPS! for the *smart people* of the world. For those not used to reading technical scientific material this could be paraphrased as, "All the stuff we believed for years was wrong. But we think we are on the verge of finding it now! We are going to start looking for *dark matter* and JUST AS SOON AS WE FIND IT, we will let you know what makes us uniquely human."

Hey gang, this is not rocket science. I mean, you can study rockets all you want and never learn what makes us human. This is animal science. And it doesn't take long to figure out that animals live in an "Alpha Wolf World" and humans live in a "Moral World." The Alpha Wolf World operates on a dominance/power system and the Moral World operates on a right/wrong system.

We have two horses. When we feed them, the BIG ONE always eats first. If we put hay in two different spots, the BIG ONE moves from place to place as he wills and the little one always eats at whatever spot the BIG ONE has vacated. When the BIG ONE lays his ears back the little one gets out of the way.

We have two dogs. When we feed them, the BIG ONE always eats from the bowl she chooses. The little one gets the leftovers. The BIG ONE barks and the little one backs off.

We have two boys. Unlike our animals, they eat quite peacefully together. But the tranquility breaks down faster than The Roadmap to Peace over who gets the biggest piece of cake, who has to do the dishes, or who gets the prize from the cereal box. And when things don't go their way, they make a MORAL APPEAL. "It's not fair!" they scream.

We live in an area with a lot of whitetail deer. They are presently in the "rut". (For non-hunters, that means it is the mating season.) For the white tail deer, SIZE MATTERS! The stronger bigger-bodied, bigger-antlered bucks fight off the younger weaker smaller bucks for the privilege of spreading their seed. The BIG ONE wins and can impregnate as many females as he is able.

In the human reproductive experience there is also something called a "rut." It is known to produce frequent headaches among the females at mating time. With humans, size does not matter.[10] But if a human male randomly impregnates numerous females we consider this a great MORAL WRONG.

The deer population in our area is greatly impacted by the weather. In years when the snow is exceptionally deep, people have reported finding deer carcasses from animals that sunk in the snow and could not escape the wolves. The wolves would simply start feeding on the entrapped live deer. A brutal and horrible death to be sure, but we chalk it up to life in the wild. In fact, that is why we call it "The Wild."

But when terrorists in Iraq cut the heads off of their unarmed, non-combatant living kidnapped victims, we experience a deep sense of *Moral Outrage!* We call them animals. We say they are NOT HUMAN to do so such a thing! Then we blame George Bush for making them mad at us.

When ants attack a beetle and carry it into their lair we give them their own television special on Animal Planet. We point out the amazing reality that an ant can carry 48 times its own weight, which is like an average man carrying a Sherman Tank. We get close-up shots of thousands of ants attacking the hapless beetle. We are all supposed to find this interesting, perhaps somewhat gross, but we are not expected to sit in judgment on the ants.

---

[10] At least this is what we men keep telling ourselves.

The Evolution of Calculators as Pick-Up Tools
Rock & Chisel ~ The "Smash & Grab" Method

When a bunch of hoodlums beat a drunkard to death, it gets covered on *20/20* and we all experience deep *Moral Outrage!* We demand justice to be done! So, the politicians begin a mid-night basketball program in the kids' neighborhood, because it's the kids that are the real victims in this scenario. Besides, any excuse to spend money in their district is good enough for a politician.

The social structure of animals is driven by the Alpha Wolf System. We count on this when training our pets. We train our dogs by convincing them WE ARE THE BIG DOG. Then they obey us.

Our dog knows she isn't supposed to be on the bed. When we are around, she never goes on the bed. But if we are in a different part of the house, she helps herself to the bed. When we come upstairs we hear her jump down from the bed. She is not misbehaving, she is being a dog. In dog life, when the top dog leaves, the next dog moves up. This is what *smart people scientists* call the PEZ Candy Dispenser[#] Theory of Animal Dominance.

In contrast to animals, the social structure of humans is driven by the Moral System. Because we want little Johnnie to grow up to be a GOOD PERSON we teach him to stay out of his older brother's room. Even when his older brother is gone for the day, Johnnie is not to go into his room and take his things. We teach him IT IS WRONG to take other people's things.

This common morality has caused some to hypothesize that the Judges who decided Kelo v. New London (which took poor people's personal property while they were at work so rich people could build a mall to increase the tax base for City Hall) were the same people documented in the tabloids as having been raised by wolves. Therefore they are still working on the Alpha Wolf System and see no problem in taking someone else's property.

At times, parents revert to the Alpha Wolf System while their kids are young. When little Johnnie asks for the 32nd time why he has to pick up his toys and the reasoned adult's logical answer just doesn't seem to be getting through, parents are reduced to tyrants, screaming, "BECAUSE I AM BIGGER THAN YOU AND I SAID SO!!!"

After having been reduced to raging ninnies parents are filled with remorse, because they are Moral Beings and they should know better. So the next day they go out and buy Johnnie *Guitar Hero* hoping he won't turn into a psychopathic killer because they yelled at him. They also know that one day he will choose their nursing home so they want him to forget this little incident from the night before.

Social scientists debate whether this moral system which defines us as humans isn't in fact the transitional form (THE MISSING LINK) that has eluded the paleontologists in the fossil record. Darwin himself said that if his theory was correct, thousands of transitional forms should be found in the fossil record. To date, the bones are quiet on this matter.

There is also the problem of TRANSITIONAL ERGONOMICS. A forelimb that is half-way between a leg and a wing is neither good for crawling, climbing, walking, or flying and as such would prove a hindrance in the battle for the survival of the fittest, reducing the ability to reproduce, and thus not evolve further.

The assertion was made years ago that "embryology begets ontology." This theory contended that gestational forms re-enact the different stages of Evolution. For example, the embryo has gill-like folds and it was assumed that these were reminiscent of when we were fish. This has since been proven wrong. They actually resemble the folds in an accordion. And "accordion" to the theory of Evolution, this is why people of every culture love to polka.

In his speech *Moral Systems as Transitional Forms*, renowned German social scientist Ido Nomuch suggested that the Moral System is still evolving and as such is a transitional form. Children are not born with moral perception, but learn it as they grow. Thus, as children, they operate and respond to the Alpha Wolf System. This is why young kids still need a good thrashing to get them to behave. Ido Nomuch suggests that "pediology begets ontology." Moral development in children reflects that we once all operated on the Alpha Wolf System.

The fact that the Evolution to the Moral System is not complete in children gives us a working example of the problem of TRANSITIONAL ERGONOMICS. The Moral Parent won't allow himself to bust his kid's chops, even though the kid deserves it. The Alpha Wolf Kid will only respond to a beating. At this point nothing is working right. The exasperated parents wind up in an argument about how to handle the misbehavior. Dad spends the night on the couch as a result of the fight, and the whole reproductive thing goes down the tubes. Scientists believe that when moral Evolution is complete, children will be born with moral understanding and peace will reign on earth.

(Years ago, via hidden cameras, there was a documentary produced on the Transitional Ergonomic Dilemma of the Wolfman. As a human he had moral anguish over his behavior. Yet when he changed to the Alpha Wolf, he couldn't help but howl at the moon and then go kill someone. It is important to remember most scientists believe that we did not evolve directly from the Wolfman. First, we became howler monkeys. About two weeks later we became humans.)

****Newsflash!****

THE VERY DAY AFTER WRITING THE LAST BRITNEY REFERENCE the following **world-shaking news** was reported on TV. I knew it was true when later that day I saw it on the cover of one of those magazines at the checkout counter that tell us every detail about famous peoples lives as if it really matters that we know this stuff.

In light of her splitting from Kevin, Britney decided to renew her friendship with Justin "Wardrobe Malfunction" Timberlake for the holidays. This raised two major questions that we all need answered. Should Cameron be concerned? And is it a good thing to reconnect with past amours at the holidays?

The question NOBODY WAS ASKING is, "If Justin can't keep Janet's wardrobe fastened, will he really be able to help Britney learn to fasten a car seat?"

Chapter 8

Opinions

*O*pinions are like noses. Everybody has one and they all smell.

This one-line witticism is intriguing on two accounts. First, its observation of the human experience appears to be accurate. Everybody has an opinion. That is how news organizations have been able to use opinion polls to go from *reporting* news to *creating* it.

> The latest MSNBC Opinion Poll indicates that 76% of women and 73% of men think Hillary Clinton is the smartest woman in the world. When asked why they thought she was so smart, nobody knew why they believed this, but 93% said they believed this "very strongly." 98% of those polled were able to name all of the Simpsons.
>
> And now, with the follow-up story to this "breaking news," the first in our 629-part series "Only Stupid People Won't Vote For Hillary in 2008," here is our

totally unbiased Keith "Impeach Bush Now!" Olberman...[11]

The second reason this witticism is intriguing is its word play on "smell." Of course, not all opinions smell. If you agree with me, then your opinion is pure genius!

This statement about opinions would remain accurate if restated about morals:

Morals are like noses. Everybody has them, and they all smell.

Everyone works from a moral basis. Everyone works from the perspective of what they believe to be right and wrong. These are our morals. Our morals are constantly giving the "smell test" to the things around us. If it smells good morally, we embrace it. If it smells bad morally, we reject it and then proclaim our *Moral Outrage!*

The *Moral Outrage! du Jour* at the time of this writing is Michael "Kramer" Richards' use of the "N Word" with some hecklers at a comedy club. By playing this over and over, bringing in psychologists, Jesse Jackson, and specialists in body language, all the important people get to express their own MORAL PURITY by criticizing the comedian. That way, we all believe that none of these "good person talking heads" ever had a racially prejudiced thought. We are, of course, expected to totally ignore Jesse Jackson's "Himeytown" comment.

---

[11] As late as August of 2008 the claim of Hillary's intelligence was still being echoed in the media. Of course, everyone also knows that Obama is a genius in his own Wright. It only took 20 years and a relentless pounding from Sean Hannity for him to reach the conclusion that Rev. Jeremiah Wright's preaching was a bit too radical for him. Now that's change we can believe in!

For over 30 years the abortion debate has raged in America, with each side claiming its stand as THE moral one. The right-to-life-people believe life begins at conception and every child has a right to live (something about the Constitution...life, liberty, and the pursuit of happiness).

The right-to-kill-people believe a woman has a right to make decisions about her own body in private counsel with her doctor. It does not matter when life begins. It is her body and her choice. The moral claim of either side is that someone's rights must be protected.

It is the Holiday Season. Every year, we battle with the question of church/state issues. Some atheists believe they have the right to zero influence of religion in the public sector. They believe a manger scene could influence people's thinking. They are correct. My cousin saw a two-foot plastic Magi in a town square manger scene and immediately went out and bought a camel. These atheists change the concept of "freedom of religion" to "freedom from religion." Others believe that religious expression within government is a part of our history (something about being endowed by a Creator with rights...as absolutely necessary to remaining a free nation). Atheists and religionists alike base their claims upon a moral perspective.

And the arguments rage over minimum wage, immigration, race, affirmative action, hate crimes, torture, women's rights, campaign funding, taxing the rich, Medicare prescription drug benefits, eminent domain, gay marriage, military draft, stem cell research, voting rights, profiling, Wal-Mart, rebuilding New Orleans, and all sorts of issues *ad nauseum*. The common link between all of these is the fact that everybody holds their position with a strong sense of *Moral Outrage!*

(Maybe the chimpanzees have it right. Think of the stress they avoid by not experiencing *Moral Outrage!* Consider family life without it.

Mother Chimpanzee: Ricky, why did your brother fall out of our tree?

Ricky: I shot him with a dart.

Mother Chimpanzee: Why did you do that?

Ricky: I am doing a physics experiment.

Mother Chimpanzee: Where did you hit him?

Ricky: In the crotch.

Mother Chimpanzee: Do you think you could hit him in the crotch if we lived in a black hole?)

****Newsflash!****

Stop the Presses! Stop the Presses!

It seems like the holidays are bringing out the longing for the good old days with Britney. Now *Fox* is reporting that she has reconnected with another old friend, Paris Hilton. Get this! Sometimes Lindsay Lohan has even been seen with them.

They partied for three nights in a row!

We have even been made privy to the idea that Britney apparently can't keep her underwear on. No wonder she can't keep a car seat in place…she can't even keep her underwear in place!

Chapter 9
Making a Buck

Because we are SO CONSISTENTLY MORAL in how we function as humans, we have arrived as a society at the inevitable intersection of the "Moral System" and "Capitalism."

America is a CAPITALIST SOCIETY. That means we believe in free markets, free enterprise, and free samples at the grocery store on Wednesdays. The free market system allows anyone with an entrepreneurial spirit to accept personal responsibility, work hard, take risks, and accumulate wealth. Once they have gathered enough wealth, the government takes it from them and gives it to others who don't work hard, take risks, or accumulate wealth. They do this because it is NOT FAIR that rich people should have more than others. The Moral System requires us to do this because WE ARE GOOD PEOPLE!

In a capitalist society, the entrepreneur is able to transform any concept into a money-making venture. Baseball cards from childhood become small business ventures. Pick up a few hand tools at a garage sale and you can immediately morph into a weekend Rent-A-Husband. Buy a camcorder, and you're a professional videographer available for weddings, bar mitzvahs, and preschool graduations.[12]

It was only a matter of time before someone realized that because we all operate on the Moral System, there must be money to be made on the fact. The basic means of making money off of the Moral System is to appeal to people's *Moral Outrage!* by identifying some injustice, unfairness, indecency, or unpunished wrongdoing, then set up an organization as the great SUPERHERO who is fighting to right this terrible wrong. Finally, ask people to send money to help fight this battle. Truly *smart people* identify a *Moral Outrage!* that can actually never be defeated. This way they can make a career of endless appeals for people's money.

The study of this intersection between the Moral System and Capitalism is called Moral-Capitalogy. Through years of rigorous study, the Moral-Capitalists have developed a MORAL-CAPITALIST NEEDS PYRAMID as to how free moral people spend their money. Absolute necessities are indicated at the bottom of the pyramid. Until all needs are met at this level, free moral people will not spend their money on the next level. All needs must be met on the second level before they will spend money on the third level, and so on with the fourth level. The Pyramid looks like this:

---

[12] Preschool graduations are important in a child's education because it is here they learn they should be promoted for doing nothing more than going to the babysitter. Then when they pass physics on Partial Credit, they are already conditioned to the idea that they deserve to pass whether they have learned anything or not.

SPIRITUAL

PHYSICAL

MORAL INJUSTICE!

ENTERTAINMENT NEEDS

The first, foremost, and foundational spending priority of free moral people is ENTERTAINMENT. This includes EVENTS (movies, concerts, sports, theater, etc., etc., etc…), ELECTRONICS (iPods, cell phones, DVD players, video games, HDTV with cable, car audio systems, computers with internet, blah blah blah…), ACTIVITIES (golfing, fishing, boating, snowmobiling, drinking, smoking, gambling, and so on and so on and so on…), and GATHERINGS (parties, girls' night out, time at the bars, la la la…).

To make sure we are keeping our priorities in order, we require weekly news updates indicating how many billions of entertainment dollars we spent on what movies at the box office the previous weekend. We wouldn't think of demanding this kind of reporting from Congress on how they spent our money the previous week or who voted for what spending. That would be boring and WE WANT TO BE ENTERTAINED! On average, America spends $537,896,332,143.98 per month on entertainment expenses.

Once our entertainment necessities have been met, we soon realize there is not enough money left over for our PHYSICAL needs such as food, clothing, housing, and medical expenses. Because we believe WE HAVE A RIGHT TO THIS STUFF, but cannot afford it, we experience deep *Moral Outrage!* We quickly realize the reason we cannot afford it is because RICH PEOPLE ARE OPPRESSING US![13]

This moves us to the second level of the pyramid.

Because we are HIGHLY MORAL PEOPLE and are aware that a GREAT INJUSTICE IS BEING DONE, we look for a solution. Lo and behold, we find the SELF-APPOINTED SUPERHEROES! If we send them money, they will go to battle for us and rectify this great *moral injustice!* Being determined to fight back against the forces of evil that are keeping us down, we send them money. Knowing that we have done our part to FIGHT THE UNFAIRNESS, we return to entertaining ourselves.

Some of the most highly skilled entrepreneurs who work the intersection of Moral System Boulevard and Capitalism Avenue are worthy of mention.

> AARP is an organization that exists to badger the government for more benefits for the elderly. Forget the fact that demographically we are heading on a path toward bankruptcy, and that over eight years ago Bill Clinton told us to "fix Social Security first." AARP is determined to keep Social Security, Medicare, and Prescription Drug benefits as the proverbial "Third Rail." Any politician who tries to touch these programs will be committing political suicide.
>
> Political Action Committees (PACs) appeal for our

---

[13] Now DON"T EVEN TRY and convince us that we could afford to pay for our physical needs if we didn't invest so much on our entertainment needs. We are Americans and entertainment is our BIRTHRIGHT. Besides, the rich have a moral obligation to give us some of their money.

money so that the right candidates will be elected to end the "Culture of Corruption," close tax loopholes, and end tax cuts for the rich. In fact, THEIR CANDIDATE IS SO ETHICALLY PURE, he will single-handedly instill the much-needed morality in Washington AND put food on our tables from off the tables of rich people.

Michael Moore is a Moral Capitalist entrepreneur worthy of special mention. He is much like Sam, the smartest of the *smart people*, who gets us to willingly spend our money on nothing. Michael is the most creative, because he never asks us for money directly but instead has effectively united the first two levels of the pyramid, ENTERTAINMENT and *MORAL INJUSTICE!* Michael makes movies that stoke our *Moral Outrage!* His movies depict as morally corrupt, the rich and powerful people who disagree with him. We willingly PAY to see his movies. We BELIEVE anything he tells us. We LEAVE the theater very angry, and yet wonderfully satisfied that someone has finally understood our *Moral Outrage!* We THINK somehow that by just watching this movie we have accomplished something. And then, because of the SUBLIMINAL MESSAGES we encountered in the movie, we go bowling. The money we spent on the movie enables Michael Moore to become richer and more powerful. He then uses our money to make another movie about the evils of other rich people who do not agree with his moral outlook.

The easiest way for Moral-Capitalist Entrepreneurs to evoke *Moral Outrage!* is to play The Size Game. The rules are easy. Simply identify a tension between two parties and always assume that the BIG ONE IS BAD and the little guy is the innocent victim. By using this approach, the ACLU is able to insure that every claim made by someone who is not a traditional American, is indicative of a little guy whose rights are being trampled.

This approach also insures that the BIG GUY IS ALWAYS THE VILLAIN!

Everybody knows Halliburton and the Oil Companies are entirely evil. Big Tobacco and McDonald's got what was coming to them regardless of any responsibility on the consumers' behalf. And the CEO of Wal-Mart is the Devil himself!

When his Entertainment and *Moral Injustice!* needs have been met, the free moral capitalist then moves on to the PHYSICAL needs of life.

In order to use his limited funds effectively, the consumer has been conditioned to use coupons whenever possible. A sense of *Moral Justification* accompanies his feeling of *Fiscal Satisfaction* when he uses his Class Action Settlement Coupon.

Because the food tasted so good and provided better value for his dollar, the consumer was a regular customer of McDonald's. But McDonald's, THE EVIL BIG GUY, had been running a very effective "SUPERSIZE IT" promotion. With this promotion, they FORCE FED their customers more food than they needed.

McDonald's hired little grandmas to work the noon day shift. These grandmas had years of experience waging PSYCHOLOGICAL WAFARE on their families in getting them to overeat by saying, "Clean up your plate because there are people starving in Africa." When the consumer heard the evil grandma say, "Do you want that Super-sized?" he immediately felt compelled to say "yes" and then proceeded to eat everything he purchased.

After years of being FORCED TO OVEREAT by the little grandmas at McDonald's, the consumer gained 329 pounds. This was a totally unhealthy situation. Along came the SUPERHERO LAW FIRM that won a 172 trillion dollar Class Action Suit against McDonald's immoral practices. Of this money, the Superhero Law Firm kept 57 trillion in cash and the rest was dispersed as coupons to the victims. They received $3 off on their next $20 dollar purchase. So the consumer, thinking he had been vindicated, bought a $20 lunch for $17. He then celebrated his moral victory of STICKING IT TO THE MAN by spending the $3 dollars he saved on a big ice cream cone. The consumer left the restaurant $20 poorer, three pounds heavier, and FEELING MORALLY JUSTIFIED over his victory.

Because this great moral victory caused him to gain even more weight, the free moral-capitalist consumer then discovered the need to SUPERSIZE his clothing. Having overspent on lunch at McDonald's, he had to shop at Wal-Mart for lower prices. This created MORAL DISSONANCE because Wal-Mart takes advantage of its employees. In a totally just world, the high school dropout who stocks the shelves should be making as much money as the Regional Director in charge of 944 stores. But because he is the little guy, the stock boy is being totally victimized by the EVIL GIANT EMPLOYER WAL-MART. The consumer managed his guilt by promising to send more money to the good people SUPERHEROES who are fighting Wal-Mart. He then made his purchase with a credit card so he could save his remaining cash for his liquor store purchases at the next stop. Somehow it just didn't seem right to purchase his beer on a credit card.

When the weekend arrived, the free moral-capitalist consumer headed to church to meet his SPIRITUAL needs. As the basket was passed, he opened his wallet and found he had ten dollars left. He gave one dollar as his "tithe." Following the service he asked the preacher why the church wasn't doing more to feed the hungry. After all, he thought, wasn't that the MORAL WAY to live?

(Moral-Capitalogists have heated debates as to which appeared first in humans, the Moral System or the Capitalist System. They have been studying chimpanzee behavior for decades to discern this development. They are trying to break the code as to what it means when chimps groom each other. Is this a "moral" behavior, whereby they help their fellow chimp, or "capitalism," whereby the most aggressive worker accumulates the largest number of bugs for his own consumption? With just a few more of your tax dollars, the breakthrough is sure to be right around the corner.)

****Newsflash!****

O'Reilly had Geraldo on tonight to discuss the problem that Britney Spears seems to be in meltdown mode. Because she is not highly educated, Bill is concerned that fame as a youngster may have led her down the path of self-destruction. Because she is living so poorly at the moment, they raised the issue that Social Services is looking into the safety of her children, and ironically, may determine for custody purposes that K Fed is actually better qualified as a parent.

NOT TO WORRY! Geraldo believes this is all a marketing scheme for an upcoming movie. Apparently she is learning how to work the paparazzi from that great marketing mogul, Paris Hilton.

Chapter 10

Grading on a Curve

Whereas "Partial Credit" got me through college physics, **Grading on a Curve** got me through high school on the subject where I was even more clueless than physics. The subject was "Social Acceptability."

I totally failed this course my first two years of high school. I was short, wore braces, and had acne. I tried to convince myself that these were merely problems of adolescence and I would one day outgrow them. Unfortunately, I haven't outgrown them. But I have added going bald and gaining weight to the mix. I did not have the athletic skills or the money to be a Jock, so I wasn't part of the "Cool Crowd." I tried to grow my hair long to look like a Hippie, but since I wasn't angry with the Establishment and thought doing drugs was stupid, I didn't fit in there either. The Greasers (a la John Travolta and Olivia Newton John) found themselves on hard times and acted more like Richie Cunningham than Fonzy. So, nobody wanted to fit in there. (Besides, I never looked good in a black leather jacket.) By default, I fell rather comfortably into the non-descript group known as "Losers."

My brother Tom, on the other hand, was at the pinnacle of the social strata. He did his best to give me a place in that circle, but alas, coolness is not transferable between siblings. Whenever I was introduced to a hot cool girl (oxymoron?), her response was predictably, "You're Tom's brother?" This question was posed with an unvarnished sense of disbelief that we were actually conceived out of the same gene pool. Our sociology teachers tried to convince us that India was the only country with a "caste" system, but any kid in high school knows they spend every day in such a place.

Some socially inept high schoolers with a moderate degree of creativity found a sense of worth by compensating with other gifts. Jimmy Oaks could suck a spaghetti noodle up his nose and out his mouth. Richard Noeluck could have easily won $50,000 on *Fear Factor* for his ability to eat any crawly thing. Some kids were artists, some geniuses, some budding comedians.

I had none of it.

I moved through my first two years of high school with a teenage angst that was off the charts. But in my junior year, I came across a kind and compassionate teacher who helped me find my place of social acceptability. This teacher brought meaning to my otherwise painful existence. For the first time in my life, everyone looked upon me as a source of goodness. All of this came about when the kind and compassionate math teacher introduced "Grading on a Curve" to the evaluation process.

In my sophomore year, the math teacher labeled me the "Class Buffoon." All of that changed the next year when I became the "Grading on a Curve Superhero." In the standardized grading method, we were graded on a straight percentage basis. By turning in scores of about 48%, I failed a lot of tests and was consistently at the bottom of the class. The other kids tended to laugh at me for my poor performance. But when the teacher introduced the new method of grading, I was instantly recognized as the MVP of Math Class.

"Grading on a Curve" meant that percentages of correct answers were not what mattered. What mattered was that there was a standard Bell Curve distribution of grades. This was just another effective means the teachers employed to camouflage the lack of learning on the students' part. Mostly C's were given out, with some D's and B's, and even fewer A's and F's. In this particular system, the entire grading standard was shifted depending on the class performance as a whole. Because I was only "slide rule smart," I was always single-handedly responsible for skewing the grading standard downward and making it easier for everyone else to achieve higher marks. Up until then the most frequently asked question was, "Will this be on the test?" With the new standard in place, the question became, "Will this be graded on a curve?"

When students knew grading would be on a curve, they would start cheering and giving me high fives. I was everyone's friend. Because I was going to score so poorly, they knew they could get by without needing to study themselves. They took turns inviting me out the night before a test, just to make sure I didn't try anything stupid like studying or something. My newfound fame and acceptance was like a narcotic. For once, my contribution was appreciated. I was SOMEBODY!

Although everyone lives according to a moral code, we all

tend to select a code that "Grades on a Curve." Our codes tend to move up and down the Morality Scale based upon the circumstances in which we find ourselves. Take the pretty young cocktail waitress who was serving the 62-year-old billionaire. After a few drinks he became rather bold and asked the waitress if she would sleep with him in return for a million dollars. Thinking a minute about what she could do with that kind of money she shyly said, "Yes." Being a hard-bargaining businessman, the billionaire asked if she would sleep with him for a hundred bucks. With great personal offense the waitress huffed, "What kind of girl do you think I am?" To which the wealthy man responded, "We have already determined that. Now we're just haggling over price."

Law enforcement grades on a Moral Code Curve. Consider the topic of profiling. When the DC Sniper was at large and people were wearing flack jackets to fill their gas tanks, the initial report was that the police were looking for a white man driving a white minivan. No one was offended they were looking for a white man. A few years later, in a post 9/11 world, TSA workers screen people at the airport based upon random drawing. Forget the reality that all terrorist acts have been carried out by 20-40 year old middle-eastern Muslim men. We have to ignore that fact and waste taxpayer money on checking little grandmas and nursing moms, because we wouldn't want to offend anyone by profiling those who actually commit terrorist acts. It's okay to look for white guys. It's not okay to look for middle-eastern Muslims.

The media grades on a Moral Code Curve. In the 2004 election, Dan Rather entered the fast lane to retirement by getting sucked into a deceptive attempt to use George Bush's military records as a means of torpedoing his re-election. This was about the fifth time people attempted the "military records ploy" against Bush. In the hubbub, what tended to get overlooked was the fact that there were groups trying to get a look at John Kerry's military records. The media never took up the cause. It was okay to use bogus documents to go through Bush's records for a fifth time, but not necessary to go through Kerry's records even once.

The NCAA grades on a Moral Code Curve. For example, the University of North Dakota has been engaged in a firestorm to keep its team name, *The Fighting Sioux*. This and other names (i.e. my Partial Credit Alma Mater "Fighting Illini") have been banned as insensitive to Native Americans. Being only *slide rule smart*, I never picked up on the ethnic slurs behind these team names. They always reminded me of the brave warriors who fought against the impossible odds of the advancing Europeans. They were fighting for their wives, children, hunting ground, dignity, way of life, and very existence. Such thoughts always gave me great admiration for the bravery of these men. It struck me as an honor paid to them that a sports team would want to emulate these qualities of strength, dignity, and fortitude. Besides, Sports Marketing is a huge business. What better way to remind everyone of the great history of these people? Eventually, when all the teams are named after animals or weather conditions we will have no cultural reminders of these great warriors.[14]

Call me stupid, but what's up with the name *The Fighting Irish* of Notre Dame? Is it even on the list of names that must be changed? This name isn't derived from the dignity of men fighting for their existence, as with the Native Americans. It comes from the idea of the Irish Temper.

---

[14] For instance, on August 6, 2007, a North Dakota radio station ran a piece on someone studying the language, culture, and stories of the last known Native American who speaks Mandan.

Brawlers. Street fighters. My paternal grandfather was a mean drunk Irishman who used to beat my grandmother. Nothing to be admired there. But the name, based on a degrading stereotype, isn't being declared "insensitive." It's okay to use the name "Fighting" when referencing the Irish, but not when referencing Native Americans.

Politicians are masters at grading on a Moral Code Curve. Politicians have machines in place which run attack ads against the other politicians. Each sends spokesmen out to the news shows to say how the OTHER CANDIDATE'S ad took the entire political process to a new low, but their ads were done with restraint. Usually their banter degrades to the level of first graders in the schoolyard.

Democratic Strategist: Your ads were MEAN SPIRITED, but we never went negative!

Republican Strategist: That's not true! You were MEANER!

Democratic Strategist: Were NOT!

Republican Strategist: Yeah! Well, you were mean first!

Democratic Strategist: Unh Uhh!

Republican Strategist: Uh Huh!

The reason for all of this is that none of the candidates have any real solutions to problems. Their goal is to get elected and maintain power. Because they have no solutions by which to distinguish themselves from their opponents, their only hope is to make the other guys look worse than themselves.

In the 2006 election, Democrats regained control of Congress by the effective use of the "Culture of Corruption" *Linguistic Charade*. Before the losers could even call for a recount, San Fran Nan was handing out political appointments to people with "checkered pasts." It's not okay for Republicans to be corrupt, but it is okay for Democrats.

The politicians can get away with this because the VOTERS grade on a curve. Think Louisiana. They vote for corrupt politicians over moral ones, as long as the corrupt politicians are promising them the things they want to hear. It somehow slips by them that if the guy is already known to be corrupt, he is probably not feeling morally compelled to fulfill the promises he made to them either.

Hollywood certainly grades on a curve. They make movies that pontificate on the moral degradation of companies that make millions off of polluting, making guns, and treating women unfairly. But they refuse any moral responsibility for making millions from films aimed at teenagers that espouse substance abuse, sexual immorality, and general rebellion towards common decency. Oh no. These movies only reflect what already exists in our society. Funny how they want their anti-corporate movies to "make a statement" but their teen immorality movies are only "reflections of society."

(*Smart people social scientists* believe moral systems that grade on a curve are part of our DNA. They theorize such systems became ingrained in our thinking through the baboons. It is theorized that two baboons were walking across a field and saw a lion charging. One pulled a pair of running shoes from his fanny pack.[15] When the second baboon

---

[15] It is believed that the incessant wearing of fanny packs by baboons created their rather embarrassing anatomical feature.

asked what his friend was doing he said he was putting on running shoes. The second baboon informed his friend that even with running shoes he couldn't outrun the lion. The first baboon replied, "I don't have to outrun the lion. I only have to outrun you."

Later that day, while the lion was still feasting on the shoeless baboon, speedy baboon was spreading his seed. The survival of his DNA was imprinted with the reality that you don't have to be good, you only have to be a little better than the worst one in the group. And so grading on a curve totally survived as the moral standard of the future.)

Chapter 11

tRNA

Morality is what distinguishes humans from the rest of the animal world. It is the characteristic that all humans share and that no animals experience. We are woefully inconsistent in our application of morality but *we never abandon it* as an operant factor in how we live.

"Mapping the Helix" cannot define our humanity. Our chemical map can neither explain "why we are moral" or "if morality is reality." This is simply because morality is outside the realm of chemistry.

Humanity is defined by morality.

Morality cannot be defined by chemistry.

Therefore, chemistry (i.e. Mapping the Helix) cannot define humanity.

Splitting the Helix

Chapter 12

The Askin' Why? Chromosome

The fact that moral behavior makes us human only served to raise the next logical question among *smart people scientists*: "How did moral perception develop?" Two theories are currently being fostered. We'll explore the first one here, then get into the second theory in the next chapter—because being a curve-setter I can only handle one at a time.

The first theory is that moral perception is a gender-based capacity. To understand this theory we need a brief understanding of gender.

Years ago, some guy floated an impossible theory that gender was determined by the planets, something about Venus and Mars. Now, I may be stupid, but I am not that stupid! If they can't even figure out if Pluto is a planet, do you really expect me to believe this planets and gender thing? Besides, I know for a fact my dad wasn't from Mars. He was from Chicago.

Gender is all about genetics, not planets. *Smart people scientists* have discovered it is completely determined by the **Askin' Why? Chromosome**.

Scientists know that our DNA exists in long double stranded molecules. These molecules align themselves into a very specific pattern. Square pegs always align opposite square holes and round pegs always align opposite round holes. The varied sequences of these four "peg and hole" combinations determine all of our characteristics.

The *smart people scientists* have given fancy names to these molecules like *adenine, guanine, cytosine,* and *thymine*. But s*quare hole, square peg, round hole*, and *round peg* would have worked just as well.

However, things are a little different on the chromosomes that determine gender. Rather than thinking of four different "pegs and holes" lining up across from one another it is easier to think of a ballroom dance line from *Pride and Prejudice*. As the two dance lines form facing each other, across from every gentleman is a lady and across from every lady is a gentleman. Throughout the dance this pattern remains constant, with the ladies on one side opposite the gentlemen on the other. With this picture in mind, it is easy to see that there is both a male and female side to the dance line.

The important distinction between these man/woman molecules is the capacities they bring. The man molecules have developed the limited capacity for *asking* the question, "Why?" For that reason they are called the Askin' Why? Chromosome. The woman molecules have developed the unlimited capacity for *answering* questions. As such, they are called the Becuz Chromosome. This chromosomal arrangement explains why women have all the answers and men, "just don't get it."

The *smart people scientists* have discovered that the Askin' Why? Chromosome is what determines gender. If the offspring receives the Askin' Why? Chromosome, it will develop into a boy. Without this chromosome, the offspring becomes a girl. For simplicity's sake, scientists rarely use the full term Askin' Why? Chromosome. Soon after its discovery it was referred to as the Why? Chromosome, and ultimately became known as the Y Chromosome by college students taking notes in Genetics class. Therefore, if the offspring receives the Y Chromosome it will be a boy and without the Y Chromosome it will be a girl.

The *smart people scientists* theorize that in pre-moral man, the Y Chromosome was rather short. As such, the male's ability for Askin' Why? was limited. Through millions of years and many random mutations, the Y Chromosome grew in length. With the increase in length came a growing capacity for Askin' Why?. An increased capacity for Askin' Why? was a favorable trait to the female gender, so these males were preferred for reproduction.

At first, males could only ask one question and this became dull and frustrating to the females. Consider the following conversation between pre-moral male and female monkeys on their first date to the local termite hill:

Male: Why do we use sticks to get termites out of their hill, but anteaters don't?

Female: Because we have opposable thumbs.

Male: Why do we use sticks to get termites out of their hill, but anteaters don't?

Female: Because we have opposable thumbs. Um, can we talk about something else?

Male: Sure! Why do we use sticks to get termites out of their hill, but anteaters don't?

Female: Because we have opposable thumbs, you moron! Would you please take me home now?

Male: Sure! Why do we use sticks to get termites out of their hill, but anteaters don't?

Female: You men are all alike! You only have one thing on your mind!

Needless to say there was no canoodlin' for that guy that night. His gene pool went nowhere.

Imagine the same female monkey on a date many nights later. She is smoking, drinking, struttin' her stuff, hopping from termite hill to termite hill.[16] She has been out with a different male nearly every night, and most of them are dead end losers on the evolutionary tree because they are limited to asking one question, "Why do we use sticks…?" We should note that for all of her riotous living, no one called her loose, or a floozy, or took the ugliest picture of her they could find and published it in the tabloids with a story about how she was having a Britney Spears meltdown. No one did this because these were pre-moral monkeys and they had no perception of right and wrong. After months and months of having the same boring conversation with every male she dated, our Little Honey of a monkey met a

---

[16] Some scientists believe termite hills evolved into nightclubs, but the fossil record is not clear on this.

pre-moral mutant male who had developed a longer Askin' Why? Chromosome. He now had the ability to ask TWO questions.

Male: Why do we use sticks to get termites out of their hill, but anteaters don't?

Female: Because we have opposable thumbs.

Male: Why are bananas good for us?

Female: Because they are loaded with potassium.

And then she thought to herself, "At last, a guy who can have an in depth conversation!"

So she chose the two-question mutant for her soul mate.

In time, the females got bored with two-question males. After millions of years and thousands more random mutations, a new three-question mutant male came on the scene and he was considered the number one catch on the block.

This process repeated itself about a gazillion times until eventually the males developed the capacity to ask questions that could only be answered with *moral* answers. The following conversation is considered to be the first conversation of the highly evolved moral beings, now officially known as "humans."

Male: Why can't we use sticks to get termites out of their hills so we can eat them?

Female: Because termites are living beings and *it would be wrong* to kill them for our own existence. Besides, bananas are good for you. Have a banana.

Male: Why are bananas good for us?

Female: Because they are loaded with potassium.

****Newsflash!****

CNN reports Britney admits she has been "far from perfect" but now she is "more mature" and will be back "better than ever."

Silly me!

I thought maturity came from learning how to handle responsibility. Instead, it comes from partying 'til you're blotto with Paris. I never knew Paris was such a role model for intellectual development and personal growth.

Chapter 13

The Teeter Tottered

Elementary physics classes always include instruction on simple machines. These include such "low tech" but ancient devices known as pulleys, inclined planes, and sticks for eating termites. It is believed by some that chopsticks evolved from the use of termite sticks. (*Chop* is Chinese for *termite*--the insect that "chops" wood.) As monkeys became more adept at using their opposable thumbs, they learned they could manipulate two sticks to pick up the termite rather than waiting for the termite to climb onto the stick of his own accord. Those early termites had not yet evolved good climbing techniques, so waiting for one to crawl on your stick could take a long time. Also, if you wiggled the stick at all, the termite fell right off. This is why the easiest way to learn how to use chopsticks is to pick up one grain of rice at a time, because our brains are hardwired for picking up one termite at a time.

Some scientists contend that the most significant of all simple machines is the lever. These men of science theorize that levers played a critical role in the whole "survival of the fittest" thing by having a great deal to do with man's Evolution as a moral being. This second theory on the development of moral perception among humans is called the **Teeter Totter Theory**.

Before moral perception had permeated the entire gene pool, it was quite common for the smaller, wimpy males to get the stuffing knocked out of them by the bigger ones. Because the females needed to know that their offspring would be protected from predators, they naturally gravitated to reproduction with the dominant males. The parents of the wimpy males experienced great *Moral Outrage!* at the abuse their kids were taking. If their kids didn't reproduce, it was the end of the line for their own genetic code. THE SINGULARLY MOST IMPORTANT PRIORITY FOR EVERY LIVING CREATURE IS THE REPRODUCTION OF ITS OWN GENETIC CODE. So, these parents introduced the lever onto the play ground. They didn't call it a lever, because that would sound like a simple machine and indicate work was involved, and no kid would touch it. They called it a "Teeter Totter" and that sounded like fun!

You can hear the difference for yourself. Picture two 600 pound silverback gorillas building a tree house:

Gorilla #1: Hey, Gus, hand me that crow bar.

Gorilla #2: Sure, Louie. Nothing like the right tool for the job! Crowbars are the best thing since termite sticks.

Okay, now imagine the same conversation with a different name for the lever:

Gorilla #1: Hey, Gus, hand me that Teeter Totter.

Gorilla #2: Sheesh, Louie, couldn't they come up with a better name for this? It sounds like a kid's toy.

The impracticality of the Teeter Totter was quickly learned by the kids. In order to operate this lever, you needed approximately equal weight on each end. The bigger males would force two or three smaller males to operate the opposite end of the Teeter Totter so they could get a ride. The smaller males soon learned if they jumped off really fast while the bigger male was in the air, he came crashing down with great acceleration. They couldn't compute the acceleration, because they didn't have *smart people calculators* yet, but it was like shooting the monkey in the crotch in physics class. You don't have to be able to reduce it to numbers to understand the outcome. Those bigger males were no longer available for breeding.

The final result? The smaller monkeys that experienced the *Moral Outrage!* of humiliation at the hands of the larger monkeys went on to reproduce. These weenie monkeys began carving their moral consciousness into the evolutionary tree. Just as the Teeter Totter needed enough weight to reach a tipping point, eventually the genetic code of the weenie monkeys' Moral System outweighed that of the bully monkeys' Alpha Wolf System. After reaching that tipping point, man always operated on the Moral System.

The Evolution of Calculators as Pick-Up Tools
Abacus ~ The "Numbers To Necklace" Method

Chapter 14

Why?

Why does it matter if Enron executives bilked their employees and stockholders out of billions?

Why does it matter if militant Islamists want to wipe out Israel?

Why does it matter if O.J. walked?

Why does it matter if our soldiers are dying in Iraq?

Why does it matter whether or not abortions are legal?

Why does it matter if hundreds of thousands are dying in Darfur?

Why does it matter whether or not gays are allowed to marry?

Why does it matter whether or not Congress raised the minimum wage?

Why does it matter if someone commits a racial hate crime?

Why does it matter if Dakota Fanning's life is ruined by playing a rape scene as a twelve-year-old?

Why does it matter if gun companies make millions off of a product that kills thousands in our country every year?

Why does it matter if polar bears go extinct?

Why does it matter if Al Gore is correct and global warming kills all life on earth as we know it?

Why does it matter if someone molests a child?

Why does it matter if a drunk driver kills a family of four?

Why does it matter if people die due to lack of medication because drug companies overcharge for their product?

Why does it matter if politicians don't have the guts to fix Medicare and Social Security?

Why does it matter if the Judiciary is being used by Democrats to get laws in place that they could never pass in Congress?

Why does it matter if Bill Clinton raped Juanita Broderick?

Why does it matter if we reject Barak Hussein Obama for President solely because he is black and his name sounds like an Arab terrorist?

Why does it matter if babies can feel pain during an abortion?

Why does it matter if poor people have their homes taken from them so wealthy people can pad the tax rolls and line their own pockets?

Why does it matter if we prohibit embryonic stem cell research even if it can cure Michael J. Fox's Parkinson's disease?

Why does it matter if Justin Timberlake disrobes Janet Jackson at the Super Bowl?

Why does it matter if the New York Times never gets hauled into court for its illegal leaks but Scooter Libby will do jail time for a process violation for a leak that was no leak?

Why does it matter if students were shot to death at Virginia Tech and Northern Illinois Universities?

Why does it matter if George Bush lied to get us into Iraq?

Why does it matter if police use racial profiling to catch criminals or even innocent people they just don't like?

Why does it matter if the gap between the rich and poor gets wider?

Why does it matter if Native Americans are offended by schools naming their sports teams after them?

Why does it matter if our government performs secret tests on our own citizens resulting in horrendous medical problems?

Why does it matter if we torture captured enemy combatants to get information from them?

Why does it matter if our soldiers torture enemy captives purely for their own enjoyment?

Why does it matter if Dan Rather used forged documents to try and sway an election?

Why does it matter if George Bush stole an election?

Why does it matter if our elected officials are tainted by money?

Why does it matter if the military keeps people locked up at Gitmo without evidence or trial?

Why does it matter if oral surgeons take advantage of their female patients under sedation?

Why does it matter if Wal-Mart totally exploits its employees and puts small businesses out of business?

Why does it matter if an atheist is offended about Christmas programs in the public schools that cram Christmas down his kids' throats?

Why does it matter if one atheist can raise a stink about Christmas and change the history of 200 years of tradition in how it is observed in America?

Why does it matter if the Ten Commandments are being forced on people in public buildings?

Why does it matter if rap music teaches young men to abuse women?

Why does it matter if the rich have tax loopholes?

Why does it matter if big media has a huge ideological bent toward liberalism which colors most of what they report as news?

Why does it matter if cheating on tests is a common practice in schools?

Why does it matter if we are outsourcing jobs?

Why does it matter if post-Katrina New Orleans was overlooked just because of its high population of poor black people?

Why does it matter if Michael "Kramer" Richards is a total racist?

Why does it matter if elite athletes are using illegal substances to enhance performance?

Why does it matter if a teacher has sex with her students?

Why does it matter if 3,000 innocent people died on 9/11?

Why does it matter if Clarence Thomas sexually harassed Anita Hill?

Why does it matter if the same women's groups that supported Anita Hill's complaint of being sexually harassed for being told a crude joke totally ignored Juanita Broderick who complained she was raped?

Why does it matter if kids are making videos of gang beatings and broadcasting them on *You Tube*?

Why does it matter if Little Jimmy rips the head off of his sister's sock monkey just because he enjoys causing her pain?

Chapter 15

Becuz'

Becuz' we are moral beings who totally believe in right and wrong.

Chapter 16

The Continental Divide

In one week I will be skiing at Big Sky, Montana. I will be foolishly tackling the slopes with guys half my age. Some aspect of the Alpha Wolf System buried deep in the *dark matter* of my DNA must be calling for me to prove that I am still the Top Dog in the Pez Dispenser.#

We will pass over the Continental Divide on our drive out west. Without the sign telling us of the Divide we would be unaware of its presence. As we cross the Divide we will undergo a slight change. Having traveled uphill, as we pass over the divide we will begin a descent. Whether it is perceptible to us or not at that moment, we will make a definite change in direction.

So it must have been as far as becoming moral beings. The change was imperceptible as we passed over our own historic Continental Divide. We transitioned from the Alpha Wolf System to the Moral System. Effectively we went from "Chemistry and Matter" as an old operating system to

"Philosophy and Morals" as a new one.

Smart people scientists tell us that humans arrived on the scene after a long process of random chemical reactions. First, this huge firecracker exploded. Then all the gases from it condensed into stars and things. The earth was really hot to begin with, but then it began to cool, creating a crust on the surface. The crust would bend and break and create little mountains and pools. Some of these pools produced amino acids, proteins, and *dark matter* that became the building blocks of DNA. Through totally random processes these building blocks became ordered into increasingly complex and self-replicating molecules. Eventually these molecules arranged themselves into single-celled animals through more totally random processes. These single-celled critters had but one thing on their teeny-tiny minds, passing their DNA on to the next generation. Through further random processes and mutations in the DNA, these single-celled things became multi-celled things. These totally random processes just kept stacking up until the multi-celled things developed organs and limbs and highly complex systems. The process of the SURVIVAL OF THE FITTEST eventually took over. Only those organisms that had mutated into being the best suited for their particular environment would be allowed to reproduce (and no nooky-nooky for anyone else).

The most important thing to remember according to the *smart people scientists* is that THESE ARE SIMPLE RANDOM PROCESSES OF CHEMISTRY AND MATTER, AND NOTHING MORE! In and of themselves they are no more significant than the chemical reactions that occur when you burn gas in your car or create gas after a good meal. We are ONLY talking **Chemistry and Matter**. "Chemistry and Matter," in and of themselves, have no moral value. No one claims that the methane produced by cows is morally superior or inferior to the oxygen produced by plants. They are simply chemical matter without moral value.[17]

At some tipping point we were no longer simply "Chemistry and Matter," but organisms who had fairly consistent perceptions of "right and wrong" as a part of their conversations. An imperceptible change of direction had taken place. We moved from chemistry to philosophy, without even knowing it. We crossed a Continental Divide of sorts and *became moral beings*.

Some scientists believe that when morality came into the picture, it created a rapid divergence in the evolutionary process. The termite hills began to be identified by their clientele. Some had names like "Pete's Place." They became the gathering grounds of the "Chemistry and Matter" non-moral types. They drank beer and laughed at jokes about how the methane from cows would probably cause Global Warming one day and kill off all the polar bears. These were called low-brow primates. Some termite hills had names like "Chez Pierre." They were the emporiums of the "Philosophy and Morals" types. They drank wine and laughed at jokes about how the levers they introduced onto the playground were going to kill off all the guys at Pete's Place. These were called "high-brow" primates. Today scientists study the brow ridges of these primate fossils to determine what type of termite hill they frequented.

---

[17] It should be noted, however, that even though they are morally equivalent, they are comically diverse. The methane from cows gets a lot more laughs than the oxygen from plants.

****Newsflash!****

Somewhere on the other side of the Continental Divide, I saw a magazine proclaiming Britney has a new boyfriend!

It's good to know she has turned her life around. What with all of her problems about car seats, and divorce, and Justin, and partying pantyless with Paris, it is so satisfying to know she has come back better than ever just as she promised. And she is going about it the right way. None of this "go back and be a mom to her kids" stuff. She went out and got a boyfriend. Because if her needs are met first then she can better meet her kids needs. It's all about you, Britney…You go girl!

Chapter 17
The Biker Philosopher

About a year after the opening of the Chez Pierre Termite Hill and Discotheque a young entrepreneur by the name of Plato thought he had a marketable idea that would ride this new wave of "Philosophy and Morals." He opened Plato's Emporium which, rather than being a nightclub, was actually a school. Because it was the first of its kind, it was considered cool and was a popular hangout for the younger crowd. The whole point was to talk about "hip" new things like "right and wrong" and "shadows and realities." The advantage to this arrangement was that Plato's Emporium didn't need a liquor license. These licenses had become very expensive and hard to acquire, due to intense government regulation after some guy named Socrates died from doing Flaming Hemlock Shots at Pete's Place. With the new current direction of "Philosophy and Morals," it became necessary for the government to create a solution to every perceived problem in society. Liquor licenses were introduced to regulate the industry so people wouldn't keep poisoning themselves.

Eventually Plato's Emporium/School of Philosophy was so popular that he invented franchising. This evolved into the modern university. In these places of "higher learning" *smart people* could gather and discuss deep philosophical things like "the meaning of life," "the suppression of women in the workforce," and "the ethics of breeding with low brow monkeys." Over the centuries, the "Philosophy and Morals" group produced great thinkers like Confucius, Sartre, and Al Gore. Everything was just ducky until the arrival of **The Biker Philosopher**.

The Biker Philosopher was not famous, but he was honest. I first encountered him among a loosely knit group of 20-somethings I used to hang with in Des Plaines, Illinois. Des Plaines is a middle class suburb of Chicago. Its two most famous claims to fame are one, that it is right next door to Park Ridge which is an upper middle class suburb that is known for being the childhood home of the *smartest woman in the world*, Hillary Rodham Clinton, "Ex-Presidential Candidate." Des Plaines is also home to the first McDonald's. That particular restaurant has been converted into a historical museum. I should say right here that I do not believe there is any truth to the rumor that the cattle futures that allowed Hillary to turn $10,000 into $100,000 overnight eventually became hamburger patties at the first McDonald's. The geographical proximity of these circumstances is as coincidental as the fact that Hillary just happened to make that $100,000 on her first try at playing high risk cattle futures. Now, that's what I call beginner's luck!

The Biker Philosopher's name was Cliff. He rode his motorcycle from the Windy City out to the suburbs. He wore his "colors" on the back of his sleeveless denim jacket. Cliff would tell stories about the other guys in his biker gang. About all I can remember at this point is that one guy in Cliff's gang was nicknamed Maggot. Cliff said this was because he never bathed. Though you might not know it to look at him on his motorcycle, Cliff was an intelligent guy who loved to discuss deep things.

Invariably, when differences in viewpoints were being expressed and our conversation began to get heated, Cliff would state his life philosophy to calm the discussion. In a classic Chicago accent, he would say, "Iddohmadduh." For those unfamiliar with the Chicago dialect, this is interpreted, "It don't matter." Terrible grammar, but straight to the point. Cliff didn't let anything bother him.

At the time, I did not understand how right Cliff was. It was only with further understanding of life that I came to realize "Iddohmadduh" is a fact. Although our entire ascent as humans was based upon the random processes of "Chemistry and Matter," we did not actually understand this about ourselves until the recent work of the *smart people scientists* from TIME magazine. We embraced "Philosophy and Morals" millennia ago, all the time believing that right and wrong were real. But when we apply The Biker Philosopher's wisdom to the *smart people scientists'* understanding that we are nothing more than the result of random processes of chemical matter, we reach an inescapable conclusion. Considering everything we have believed about "Philosophy and Morals" and "right and wrong," we can only conclude, "Iddohmadduh". The fact is, WE DON'T MATTER. Nothing matters. Nothing matters because everything is simply "matter" going through various chemical reactions.[18]

---

[18] Some *smart people pop music lyricists* are currently doing research to see if the phrase "Iddohmadduh" can be traced back to "Hakuna Muhtata" ("no worries for the rest of your days") from Timon and

Pumba of *The Lion King*. They have found the markings on the cave walls of Rafiki, the Baboon Prophet, and are comparing them with the Chicago graffiti, "where the words of the prophets are written on the subway walls, tenement halls" (Simon and Garfunkel, *The Sounds of Silence*). Results will be forthcoming.

Chapter 18

The Chemical Continuum

Which of the following chemical reactions is more moral?

Igniting the gun powder in a 4th of July firecracker or burning a scented candle to make a room smell nice?

Dissolving the grease on the dinner dishes with dish soap or removing stains in your clothing with an enzymatic detergent?

Producing methane through bovine digestion or producing oxygen through photosynthesis?

If you are still pondering the answers to these questions you have a highly inflated view of your own intelligence. These are trick questions. No right answers exist, because chemical reactions are not moral in and of themselves.

Consider the following continuum that goes from the Big Bang to the Big Bong. The Big Bang represents the hypothetical explosive chemical reaction which is basically the beginning of things as we know them according to *smart people scientists*. The Big Bong represents a hypothetical case of a father not only smoking dope in front of his small children, but actually teaching them how to take a toke for themselves in order to induce particularly pleasant electrochemical reactions in their brains. If everything is ONLY "Chemistry and Matter," at what point do we identify something as *Moral*?

> THE BIG BANG *begat* PROTEIN FORMATION *begat* SINGLE CELL METABOLISM *begat* MULTIPLE CELL METABOLISM *begat* INVERTEBRATE METABOLISM *begat* VERTEBRATE METABOLISM *begat* MAMMAL METABOLISM *begat* PRIMATE METABOLISM *begat* NEANDERTHAL METABOLISM *begat* HOMO SAPIEN METABOLISM *begat* BIG BONG BRAIN CHEMISTRY

I wrote this 'continuum' and question about where morality kicks in about two weeks ago, with no real idea on how to answer the question. But that all changed yesterday. Kudos to the entertainment industry! Those pillars of moral stability and unerring discernment have allowed me to at least get Partial Credit on my question. Last night I discovered the show *Men in Trees*. With a title like that I had to watch it. It didn't disappoint me!

The show was a commentary on men's sexuality from a woman's perspective. The men are about two mutations away from the termite hill monkeys on the prowl at *Pete's Place*. The women are highly sensitive, honest, and morally grounded in their sexual expression.

Much of the show takes place in a bar. One gets the

impression that the termite hill evolving into a bar has been more successful than male monkeys evolving into moral men. This is because, although the show allows for differences among individual men, one thing is clear about ALL MEN! ALL MEN! believe sex with no strings attached is the best sex ever. ALL MEN!

The gist of the show is the angst felt by the evolutionarily advanced, highly moral, sexually sensitive women left to struggle with the mating process among the less evolutionarily developed, low/non moral, sexually animalistic men available to them. The representation of men by this show is that they have evolved to only about the three-question capacity (Why do we eat termites…etc).

The highly moral women in the show are straddled with the burden of helping morally undeveloped men deal with their problems. These problems include being a father to another man's child, not being able to father a child, wanting to risk fathering a child without taking any responsibility for the child, and "Mannin' up!" at all costs when you have fathered a child.

At the end of the show, we are given the conclusions of Marin, the lead woman. While allowing for the fact that men are different as individuals, she affirms the value of "good men." This is a *moral* statement. Inferred is the idea that good men are superior to bad men.

So here is what we learn about morality on The Chemistry Continuum, inferred from the wisdom of the *Smart People Writers* of television comedies:

Women are more advanced than men.
Women are more moral than men.
Women get to define morality.

Women get to define morality for and about men.
Men are barely out of the primate stage on issues of morality. (That is why they call it *Men in Trees!)*
Men are still wrestling with figuring out morality.

Therefore, somewhere after the primate stage and during the Homo Sapiens stage, "Chemistry and Matter" magically took on the robes of morality! Exactly when that was will be determined by females, of course, because males are clearly too obtuse to know any better.

****Newsflash!****

Uh-oh! Problems in Happy Land.

The tabloids are talking about drug abuse and the possibility of rehab for Britney.

Maybe this "get a new boyfriend" didn't solve all her problems after all. Rats! I was SOOO CONFIDENT that a new man in her life would totally solve all of her problems.

Or, maybe she just needs to find someone more advanced than a "three-question" model.

Chapter 19

No Partial Credit

If only reality were as kind as the *smart person grader* in physics class. I never had to really get the answer right in that class. I only had to go sorta' far enough in the right direction and they would give me enough credit to pass.

I want so badly to believe that the "Philosophy and Morals" we embrace are real. But if the *smart people scientists* could correctly prove that we are ONLY "Chemistry and Matter," then why would my matter, matter?

If "Philosophy and Morals" were even a little true, maybe I could get Partial Credit. But, if we are merely the culmination of eons of random chemical reactions, if we can be defined entirely by our DNA, then we don't matter any more than a rock.

And there is no Partial Credit on this issue.

When the *smart person physics professor* shot the monkey

in the crotch it was funny. It's not intelligent humor, because the entire class laughed uproariously at the event. If the dart totally missed the monkey and both dart and monkey simply fell to the floor, it would not be funny. It would just be a meaningless physics demonstration that looked rather stupid. The reality of the dart hitting the monkey's crotch is what made it funny. You could boldly proclaim, "Even though the dart missed the monkey I am *calling* it funny!" But when someone phones it in to that place on *TBS* television where people measure exactly how funny something is, it will get a total zero. Just because you call it funny doesn't actually make it funny.

And just because we say something is moral doesn't make it moral.

Especially if there is no such thing as morality.

Nope, no partial credit here.

Chapter 20

The Ultimate Moral Dilemma

I was traveling in Minneapolis and caught a radio talk show during the heyday of *Air America*. I think their heyday fell on a Thursday, but I can't say for sure. Whatever day it was, I was fortunate enough to be in a big city that day and actually got to hear some of their historic programming.

"Historic" is the only way to refer to this particular entrepreneurial enterprise, because it was pretty much history before it got started. The program I heard appealed to people's *Moral Outrage!* and paraded itself as the antidote to the lies being perpetrated by that scum of the earth Rush Limbaugh. The announcer punctuated the airwaves with numerous and colorful invectives. I am positive she convinced every one of her eight or nine listeners that she was the source of all truth and would be the savior of the culture by wresting it from the clutches of the evil conservatives. Trying desperately to fill air time while waiting for callers, the announcer wanted to go to the phones, but no one was there. At one point she said, "We had a caller, but they hung up." Even a "slide rule smart" person like me knew this was a radio program that was bound for the ash heap of history.

Not to worry, though. The Democrats, being the standard bearers of all morality, are considering bringing back the Fairness Doctrine. That means for every minute of conservative broadcasting time there has to be a minute of liberal broadcasting time. The conservative viewpoint will have to operate like any other business by producing a positive cash flow for its product. The liberal viewpoint, on the other hand, will simply be given airtime at the conservative viewpoint's expense. This is deemed moral, good, and "fair" by those who cannot compete in the marketplace of ideas.

Finally, someone called in.[19]

The caller was a mother who proclaimed with pride that she was sending her son to a summer camp for atheists. She mentioned two specific components of the camping program. One was archeological digs to examine the fossil record and learn how we descended (ascended?) from monkeys. The other was class time which would teach the campers how to be moral human beings and good citizens of the earth. I hope for this woman's sake that her son was no brighter than the people who started *Air America*. If the boy was even one IQ point higher than the AA bunch he was going to quickly confront his mom with the Ultimate Moral Dilemma. If we are only a unique arrangement of "Chemistry and Matter" that has evolved through totally random processes, how do we ascribe moral value to our

---

[19] This call provided some much needed comic relief. The caller's moral arrogance as a parent was every bit as humorous as the radio host's self-proclaimed moral rectitude.

actions or persons when we are of no more consequence in the universe than a rock?

Imagine this conversation:

Mother: Jimmy, what did you learn at camp?

Kid: I learned we came from monkeys.

Mother: And where did the monkeys come from?

Kid: They came from less evolved mammals.

Mother: And where did the mammals come from?

Kid: They came from amphibians?

Mother: And where did the amphibians come from?

Kid: They came from fish.

Mother: And where did the fish come from?

Kid: They came from single-celled aquatic animals.

Mother: And where did the single-celled aquatic animals come from?

Kid: They came from primordial soup.

Mother: And where did the soup come from?

Kid: It came from the minerals in the earth's crust as it was cooling.

Mother: And where did the minerals come from?

Kid: From the Big Bang.

Mother: My! Aren't you smart! What else did you learn?

Kid: We learned that we should be good people.

Mother: Yes, you should. I am so proud of you learning these things. You are such a good boy!

Kid: Mom, can I ask a question?

Mother: Sure, anything dear.

Kid: If we are only the result of random processes of "Chemistry and Matter," and if "Chemistry and Matter" are not moral entities of themselves, how can we label some actions as morally good and some as morally bad? Who knows for sure what is good and bad? Do good and bad actually exist? How do we know? How do we know what we call "good" is really good?

Now Mom is confronted with the Ultimate Moral Dilemma. As a moral being, she wants to be honest. She sent her son to camp in order to learn morality. But if she is going to be honest, she has to admit that *since we are only "Chemistry and Matter," there is no basis upon which one may make a moral assertion.* There is no way to know for sure what is right and wrong. Is she to be morally consistent and tell her son the truth, that they really don't know what they're talking about? Or should she be morally inconsistent and fabricate an answer to make him believe there is a basis for morality?

So, as quickly as a parent being asked where babies come

from, she changes the subject.

Mother: Hey, do you want to go to the zoo tomorrow? I hear they have a great monkey exhibit there!

****Newsflash!****

A sincerely sad note in the news at the loss of Anna Nicole Smith.

She was blonde, beautiful, sexy, wealthy, famous, loved to party even though she had a baby to care for, abused drugs, hung around with poor company, and lived in self-destruct mode while her "friends" stood by and did nothing.

The public is awaiting toxicology reports to show the cause of death.

A book and movie deal are sure to be forthcoming.

There is no truth to the rumor that Britney is considered a "lock" on the lead roll just because she is already living this life herself.

Chapter 21

Presuppositions

She was sweet. She was pretty. She was the love of my life. And she broke my heart the day she arrived in class wearing an engagement ring! To a sophomore in high school who did not meet the social requirements of coolness, athletic ability, social standing, or academic achievement (grading on a curve didn't come until the next year) it was not hard to fantasize that Miss Lutz, the pretty young English teacher, would wait for me and we would spend our lives together. Somehow, I was living with this crazy presupposition that, when I graduated from college, I would return to my high school and Miss Lutz would not have aged. We would begin dating and she would accept my proposal for marriage. But the reality is you cannot catch up in age to someone. As I aged, so would Miss Lutz.

So it is with morality.

If we are only "Chemistry and Matter," regardless of how complex our molecular arrangement becomes, we can never "catch up" with morality. In fact, if we are only

"Chemistry and Matter," the entire discussion of morality makes no sense. We must begin with the presupposition of morality in order to discuss it as part of our experience. But just as Miss Lutz would tell us, "You may not use the word to define a word," *so we cannot presuppose the existence of morality in order to prove morality.* "Chemistry and Matter" cannot produce morality in and of themselves, any more than my dreaming about Miss Lutz marrying me would ever cause her to fall in love with me.[20]

---

[20] Sometimes I like to believe the whole "crush on Miss Lutz" thing simply proves that I was the classic MAN BEFORE HIS TIME. Nowadays, it almost seems common that pretty young teachers monkey around with their students. And why not monkey around with minors if we are nothing more than "Chemistry and Matter"?

Chapter 22
Wobbly Stars

Did you know some stars wobble?

I did not know this until recently. I learned that s*mart people scientists* are able to see them wobble. I don't know for sure how they do it, but I think they launched a *really powerful telescope* into space some years ago that sends back pictures. It is called the Wobble Telescope.

The scientists have concluded that the wobble is being caused by some invisible mass with so much gravitational pull it is impacting the light from the stars. They can't see the mass, but they assume its presence because of the wobbly light from the stars.

Some of these smart people scientists can see EVERYTHING so clearly through their telescopes, they can prove beyond any doubt we are only the result of a long process of chemical reactions. In fact, these *smart people* know so much about everything they can say with ABSOLUTE CERTAINTY that there is no intelligent design behind our existence. Now, a watch, a building, or a car, all presume a designer. But human beings, which are magnificently more complex, are here solely as the result of unintelligent random processes. I think these people must have REALLY BIG CALCULATORS to be able to prove this.

But being only "slide rule smart" myself, I am just stupid enough to ask the really dumb questions. For instance:

> Why has Condoleezza Rice never been touted as an empowering example for either blacks or women by any ethnic or women's organization?

> Why do celebrities claim they love their fans but never invite them over for the holidays or for parties?

> Why do we not conclude there is a Moral Designer that is the source of the moral component common to all of our experiences? Perhaps, even though we can not see this Moral Designer, its presence can be assumed by the existence of the moral construct in the "Chemistry and Matter" types known as humans, just as the wobbling in stars is created by an unseen mass impacting the light from the stars.[21]

---

[21] I have to be honest that I am not the first to ask this question. A large group of people dedicate their lives to contemplating the existence of a Moral Designer. They live in monasteries. They are called monks. They produce great wines and invented peanut butter cups. That is why some people believe they are the direct descendants of Reese's Monkeys.

Chapter 23

Pablo, the Mayor, and George

A city the size of Chicago has many points of interest.

It has interesting history like the Great Chicago Fire and the St. Valentine Day's Massacre. (The father of my childhood friend Wayne lived in the neighborhood and was there the day the massacre happened.)

It has interesting landmarks like the Water Tower (left over from the fire), the Sears Tower, Wrigley Field, and the Prudential Building.

It has renowned villains like Al Capone, John Wayne Gacy, Richard Speck, and the guys who work at City Hall.

It has politics like no other city.

It has excellent museums like the Science and Industry, Natural History (with authentic and amazing shrunken heads), Shedd Aquarium, and Adler Planetarium.

It has great sports teams like the Blackhawks, Da' Bears (who played really well in Super Bowl XLI for 14 seconds by returning the opening kick-off for a touchdown and never scoring again after that), the Bulls, the White Sox and the ever lovable Cubs, who are the only team in baseball with *fans so "slide rule stupid"* that they interfered with the game and totally reversed the Cubs momentum[22] in the 2003 playoffs.

It has world class arts at the Opera House and Art Institute.

It has ethnic neighborhoods like Andersonville, Chinatown, and the South Side.

It has nightlife on Rush Street and Navy Pier.

It has Oprah Winfrey and the Magnificent Mile.

It has millionaires along the Gold Coast and drunks on skid row.

It has historic places of service like Pacific Garden Mission. Known as "The Old Lighthouse," thousands of down-and-outers have found hope and a helping hand there.

It has great educational opportunities like Moody Bible Institute, Northwestern University, DePaul, Loyola, and the University of Chicago.

It has great food like Italian Beef sandwiches, Vienna Beef Hotdogs (if you haven't had one you haven't had a real hotdog), and Lou Malnotti's pizza (the thick crust Italian sausage is to die for!).

---

[22] Did you notice how *sports smart* I sounded by injecting this Linguistic Charade here?

And Chicago has THOUSANDS *of very interesting people*!

George was not one of them.

George was my wife's grandfather. He was born in 1901 and spent his entire life on Chicago's north side. He raised his two children in a two story row house purchased for $4,000. He worked for Singer Sewing Machine during the week, but his real love was being a musician on weekends. He played a gold plated tuba and the stand up string bass. Once he played with Tommy Dorsey. He knew and loved the city. He also had a habit of becoming talkative after a few drinks.[23]

Frequently, by the time we arrived for family gatherings, George had already shnockered back a couple. Just a few more martinis and the words would flow. One day, while the conversation unfolded, George and I got to talking about *The Picasso*.

The Picasso is an original sculpture donated to Chicago by the great artist, Pablo Picasso. It can be seen at the end of the movie *The Blues Brothers*. The sculpture created quite a stir when it was unveiled because *no one knew for sure* what it was. I don't think there is any truth to the rumor that some astrophysicists thought it was supposed to be Pluto. When you look at Picasso's paintings, you see he puts body parts anywhere. His sculpture was just as confusing.

---

[23] This is a whole lot better than my own Fighting Irish grandfather from Chicago. He tended to throw fists rather than words when he got drunk.

Tuba-playing George put forth a fascinating interpretation of the sculpture. George said the sculpture was a joke played by Picasso on Mayor Richard J. Daley. Mayor Daley (the father of the more recent mayor, Richard M. Daley) was considered the last of the Big City Bosses and will always be remembered for his "shoot to kill" order given during the Democratic National Convention.[24]

George interpreted *The Picasso* as a caricature of the Mayor. He said the Mayor always looked down his ample nose at people when he gave orders and talked to the press. He mumbled his words, because he didn't open his mouth very far when he spoke. He wore pinstriped "power" suits. Each of these elements can easily be seen in the sculpture if you interpret them this way.[25]

I asked George why he didn't send his interpretation to the newspaper to clear up the confusion. He simply said if he did send this in, he knew the police would be at his door the next day and his life would never be the same. He had been watching Mayor Daley long enough to know you didn't cross The Man. So, Grandpa George took this knowledge to his grave.

Personally, I think he was right. I cannot see the sculpture without seeing Richard Daley in a necktie and pinstriped suit, looking down his long nose and mumbling to the press.

I guess, because George thought it was Daley and because I

---

[24] I personally did volunteer work for the Humphrey campaign during that convention. I watched the protesters in Grant Park giving their speeches and saw a military jeep pass by with a machine gun mounted on the back. Pretty exciting stuff for a boy of 13!

[25] Some may find George's theory about the sculpture curious. This is not to be confused with Curious George, the cartoon monkey.

think it is Daley, then IT MUST BE DALEY!

WRONG!!!

There is only one person who can say for sure if that sculpture is a caricature of Richard J. Daley. His name is Pablo.

Everything else is *dark matter*.

Unless Pablo tells us, we can never know for sure. And he's no longer talking.

The same holds true for morality. We all interpret circumstances, actions, and concepts as moral or immoral. But *none of us knows for sure* if something is moral or immoral unless there is a "Pablo" (Moral Designer) who is the author of morality and who defines morality for us. Wouldn't it be great if, unlike Pablo, he were still talking?

Chapter 24

How to Feel Smart

"One smart fellow, he felt smart...
Two smart fellows, they both felt smart."

This little tongue twister is inherently self-contradictory. The topic is "feeling smart." But if you say it really fast a couple of times, pretty soon you will be doubled up in laughter. We know intelligent humor doesn't make us laugh, so you wind up feeling stupid even though you are discussing feeling smart.

Although my school days were a relentless tutorial on the fact that I was "slide rule stupid," Miss Yell tried to challenge that belief. Don't let her name fool you. She was young, pretty, and very nice. We were her first class. That accounts for her idealism by which she believed I could amount to something one day. I wish she had been correct. I was only in third grade so I didn't have the same fantasies about marrying her like I did with Miss Lutz. Besides, I was still in love with Debbie Prysi from first grade.

Miss Yell told us that there was no such thing as a stupid question. This, too, is inherently self-contradictory to me. Doesn't the fact that I am asking the question indicate my stupidity? But I decided to take Miss Yell at her word, because she had a push button calculator. So every time I wanted to feel smart, I would ask a question, and then I would know I wasn't being stupid. Unfortunately, the next time I had to *answer* a question I was usually wrong and felt stupid again.

Ever since third grade, in order to cover up the reality that I am pretty stupid, I ask a lot of questions. Psychologists call this "compensating." The questions that cause me particular distress are the questions that seem to be self-contradictory. So I'll ask them, and I'll feel smart. Hopefully, you won't catch on to how "slide rule stupid" I really am.

WHO CREATED MORALITY?

People who are too smart to believe in God like to ask those dummies who do believe in God the question that to them has no logical or rational answer, "Who created God?" The dumb people always retreat into the corner with the same answer, "Nobody created God. He is eternally self-existent." So my question is, "Who created morality?" If we are only a unique arrangement of "Chemistry and Matter," when did these things develop a moral component? Doesn't discussing morality assume its reality? Who created it? Don't we have to conclude it is self-existent? If morality is self-existent, why is the idea of a self-existent Moral Designer seen as impossible?

ISN'T LIFE MEANINGLESS?

Stars burn out. Candles burn out. Forest fires burn out. Nobody cares. So what if a life burns out, or is wiped out in a war, murder, or catastrophe? It is inevitable anyway. Nature constantly induces death, yet we don't react with horror and aversion. So, why do we place a moral value on life, especially on human life? Why do we act like war, murder, or catastrophe are terrible things? "Chemistry and Matter." Minerals and water. Nothing more. Iddohmadduh.

WHAT'S SO UPSETTING ABOUT CHRISTMAS AND THE TEN COMMANDMENTS?

Why does it matter if the values and worldviews of the Christian traditions are known in our public life? Why do they generate such *Moral Outrage!* from those who know them to be myths? They certainly don't react the same way to Santa Claus and the Easter Bunny? Who cares what foundation we build our morals on? Our public restraints (laws) are based on nothing more than opinions, if we are only the result of a long evolutionary process. Why can't all opinions be heard and discussed? If the non-Christmas opinions are so superior to the pro-Christmas opinions, won't this be clear for everyone to see? If not, who cares? It's only about gaining power. I understand the anger at being thwarted from gaining power. We are all inherently selfish and want our own way. I don't understand the sense of *Moral Outrage!* at allowing the expression of historic religious foundations, if life is ultimately meaningless anyway. Really! Who cares? "Chemistry and Matter." Minerals and water. Nothing more. Iddohmadduh.

WHY WORRY ABOUT GLOBAL WARMING?

What if global warming kills all life on this planet as we know it? So what? Pluto doesn't have life on it. (Or at least that is what the *smart people* are telling us for now.) Why does it matter if life continues on Earth? Why does it matter if humans survive? "Chemistry and Matter." Minerals and water. Nothing more. Iddohmadduh.

AREN'T ATHEISTIC EVOLUTIONISTS SELF-CONTRADICTORY ABOUT THEIR MORAL INSISTANCES?

Atheists insist that we teach only atheistic Evolution in the classroom. Why does it matter if students learn there are some less enlightened individuals who believe there is a Designer behind all of this? Why the *Moral Outrage!* that something less than truth might be discussed? Can they not simply explain that Evolution has selected for some people who believe in a Designer as one of the many variations in the great cosmic game of Mr. Potato Head? We all just die in the end anyway. Who cares what we believe in the meantime? Why such an intense demand for TRUTH when there is no ultimate morality anyway? "Chemistry and Matter." Minerals and water. Nothing more. Iddohmadduh.

ISN'T "*MORALITY IS ULTIMATELY MEANINGLESS*"[26] THE ONLY NON-CONTRADICTORY POSITION THAT AN ATHEIST CAN HOLD ABOUT MORALS?

If the atheist insists that we are here solely as the result of eons of random processes known as Evolution, then mustn't he or she conclude that meaning, purpose, and morality can only be embraced with the understanding that

---

[26] Read "Vestigial Organ" if you want a new *smart people word* for this empty morality.

they are ULTIMATELY meaningless concepts? The atheist certainly can and does embrace them, but honesty (a moral position) requires the admission that they are of no real substance or foundation. "Chemistry and Matter." Minerals and water. Nothing more. Iddohmadduh.

WOULDN'T A "MORAL SYSTEM" BE AT A NATURAL DISADVANTAGE TO THE "ALPHA WOLF SYSTEM" AND CONSEQUENTLY DIE OUT FROM NATURAL SELECTION?

Take Iraq. Nobody argues with the fact that our military could easily decimate any of the enemies we face there. But we are required to act "morally." No wiretapping. No torture. No mistakes about whom we kill. No collateral damage. No humiliating captives. No profiling. Talk to our enemies. They are nice people. Explore all of our own moral failures and put them on TV for all to see how terrible we are. Call our troops mercenaries and occupiers.

Our enemies operate under no such moral constraints. Kill innocent people intentionally. Cut off heads of living people. Torture. Maim. Anything goes.

Who did we hear was losing? We were. We were being told we can't win. We can win. We just can't win exercising morals against an enemy who interacts with us entirely according to the "Alpha Wolf System." In this case, the moral-less were surviving.

So we initiated the "Surge", a supremely "Alpha Wolf System" approach to the problem. And lo and behold, we are winning. You, of course, may not know this because as soon as we started winning, reporting on the war by our major networks dropped off precipitously. Apparently, the only truly moral position that the networks find worthy of reporting is when we are losing at war. Some claim this lack of reporting is because the networks will not report anything that makes George Bush look successful. But that would lead one to conclude that the media had an agenda beyond simply reporting the truth. But don't you believe it! These paragons of moral rectitude could never be motivated by anything other than journalistic integrity.

But even if they are not honest with us, who cares? "Chemistry and Matter." Minerals and water. Nothing more. Iddohmadduh.

IS MORALITY REAL?

Consider rape. It is generally accepted that rape is an immoral act. I would assume the feelings of humiliation, violation, anger, fear, powerlessness, revulsion, and retaliation are all intensely felt, and these are only a few of the feelings processed as a victim.

But *smart people scientists* can reduce the victim's feelings to particular places in the brain and to the chemical and electrical pathways that produce them. Why is it immoral for the rapists to induce these chemical/electrical pathways in their victims? We induce chemical/electrical pathways in other areas of life, and we don't consider them immoral. For example, we don't say it is immoral for the rapist to induce the chemical/electrical pathways necessary to drive his car to the place of perpetrating his crime.

Must we conclude that our standard of what is moral or immoral is nothing more than the codification of our emotional responses to particular behaviors?

"Chemistry and Matter?" Minerals and water? Nothing more? Can we honestly say, can we possibly live with, and do we really believe, "Iddohmadduh"?[27]

---

[27] Finally, inquiring minds want to know...
When they sent the first monkey into space, did he wear those funky space diapers made so popular during the 12,000 mile non-stop car ride of the crazed female astronaut?

****Newsflash!****

I finished writing on a Saturday what you've read so far.

The next Monday my February 12, 2007 *Newsweek* arrived, and *I am not kidding*, Britney and Paris were on the cover!

Coincidence? I don't think so!!

For the first time ever, my copy arrived a day early. *Newsweek* always arrives on Tuesday. They must have figured this was an especially important issue to get it here a day early. The cover story was "The Girls Gone Wild Effect." It wrestles with the *moral influences* such celebrities are bearing upon our young girls.

This is the article's most significant revelation:

> "For better or worse, it may also be that they now feel entitled to dress as crassly as they choose, date unwisely and fall down drunk, *the way men have since the dawn of time.*" (Yeah, I am still the guy with the whole "emphasis" thing.)

I am thoroughly delighted to know that the authors, Kathleen Deveny and Raina Kelley, with the help of the other ladies Karen, Susannah, Anne, Julie (and Jamie?) have verified by this sweeping statement that we males truly haven't progressed morally since we were *Men in Trees*.

But that leaves me wondering. Does Britney's behavior (mimicking the men who are still in trees) reflect a reversal of the evolutionary process? If so, does this mean her kids are returning to the monkey stage? And if that is happening, is it possible that her grandchildren will be able to star in *Planet of the Apes 37* without having to wear costumes?

Of course, my question assumes that Britney will first have to learn to use a car seat so she doesn't kill her kids before they can make those grandbabies.

Chapter 25

Dark Matter

A number of weeks ago, Britney began hiding out in rehab. Without regular *Newsflashes* to post, I figured the average entertainment-oriented American reader would get bored rather quickly with anything more I had to write. Therefore, I decided to turn my attention to the logistics of publishing. I will finish the rest of the work when Britney returns to public life.

While on my writing hiatus, some major events occurred. Each event relates to the question behind my quest in this work.[28] A few of them merit mention here.

First, we must note that *Moral Outrage!* has CONSISTENTLY filled the news.

Second, we will observe that through the ongoing and

---

[28] What was that quest again? To understand who is right and how we know they are right concerning all the *Moral Outrage!* that constantly surfaces in our culture.

widespread process of "Grading on a Curve," moral standards have been INCONSISTENTLY applied in addressing all that outrage.

Finally, even a casual observer of all this outrage will readily notice that because we are a Nation of Laws, certain legal processes must be followed when *Moral Outrage!* erupts. The order of the process may vary, but each step of the process must be executed.

Videotape, audiotape, or a personal confession of the offense is released to the news agencies. These news programs run the offense every hour for about two or three days. In a very a short while the incident is replayed about 32,000,000 times until the entire nation can recite it from memory. Now that it has become such BIG NEWS, it requires Commentary in addition to Reporting during each hourly news cycle.[29]

The Perpetrator is then hauled kicking and screaming into the Supreme Court of Public Opinion. Now that his or her life has been destroyed by the Reporters and Commentators, the Perpetrator has no recourse but to go before a Judge and beg for forgiveness. The Judge then hands down a decision.

---

[29] It should be noted that one must first be "qualified" to comment on the *offense du jour* which has generated the *Moral Outrage!* All people wishing to comment must demonstrate their qualifications to speak on moral issues in the following manner: Regardless of one's opinions on the offense, political correctness requires that one describes the offensive behavior as about the WORST THING EVER SEEN ON THE FACE OF THE EARTH! One must first express their personal *Moral Outrage!* with the use of words like *despicable, disgusting, outrageous,* and *unforgivable.* Having hurled such epithets, the Commentator establishes himself as *Morally Superior* to the Perpetrator. This also leads us to assume that their personal lives operate on a PERFECT MORAL STANDARD and they are therefore "qualified" to sit in judgment on others' behavior.

Miss America and Alec Baldwin

This interesting scenario began when **Miss America** confessed to abusing alcohol with Miss Teen America. Being found guilty in the Court of Public Opinion, she went before Judge "The Donald" Trump and pleaded for mercy with tears and sincerity. Judge Trump handed down the decision that she should be forgiven, provided she went into rehab. That excited the *Moral Outrage!* gene in Rosie, and the two were off to the races. Notice their CONSISTENT *Moral Outrage!* as they faced off in the media:

Judge Trump: What Miss America did was despicable and disgusting. I have decided to pardon her provided she enters rehab.

Rosie: Pardon her?! What she did was despicable and disgusting. The only reason that old geezer pardoned her is because she wiggled her little butt for him.

Judge Trump: What Miss America did was despicable and disgusting, and the only reason Rosie is upset is because she has a fat butt.

Rosie: What she did was despicable and disgusting, and Donald Trump should be able to buy better hair with all his money.

Judge Trump: What Miss America did was despicable and disgusting, and Rosie is a liar and a sad and pathetic loser and she will be off *The View* in no time because Barbara Walters can't stand her.

And so the insults were hurled back and forth, which really

made Miss America happy, because people got so busy watching The Donald and Rosie, they forgot that her drinking is what started the whole thing.

A few weeks later, **Alec Baldwin** reamed out his daughter on the family answering machine.

Bad call, Alec!

NO ONE KNOWS HOW, but that recording got released to the press and played over and over and over. (Since when did parents reaming out rude teenagers become "national news"?) Soon Alec found himself skewered in the Court of Public Opinion and had to go before a Judge to seek forgiveness. He went before Judge Rosie, because he knew she would understand what it means to undergo criticism in the public eye. Actually, he was just looking for a soft shoulder to cut him some slack. He found just such a thing in Rosie's Court, a.k.a. *The View*. Notice the CONSISTENT *Moral Outrage!* accompanied by the INCONSISTENT Moral Standard when measured against the Miss America incident.

Judge Rosie: Alec, what you did was despicable and disgusting, but I forgive you because we all get upset with our kids at some time. Shoot, I was so mad the other week I gave my own kid a swirly.

The Donald: Forgive him?! What he did was despicable and disgusting and is nothing short of child abuse. The only reason Rosie forgave him is because this is the only way she will ever get a man to appreciate her.

Judge Rosie: What Alec did was despicable and disgusting, and I wish one of Trump's buildings would collapse on him like Building 7 of the World Trade Center, which was the first time in history fire ever melted steel!

The Donald: Alec was despicable and disgusting, and if Rosie had half a brain she would understand fire is what is used to form steel in the first place.

Judge Rosie: What Alec did was despicable and disgusting, and The Donald is too dumb to understand that Bush personally planted the dynamite in Building 7 which made it collapse, and I am leaving *The View* because it just isn't worth $10,000,000 a year to sit here and spout my ridiculous and totally unfounded assertions.

Hmmmm...forgiveness for Alec, but not for Miss America. Tell us about those high, moral, and consistent standards again, Rosie?

Duke La Crosse and Rutgers Basketball

In the run up to the '06 elections, our nation was treated to the story of three Duke University La Crosse players accused of rape by a black female stripper. Apparently because they were "Dumb Jocks" these three players did not realize that they were required to go before a qualified judge (someone who declared what they did was "despicable and disgusting") and beg forgiveness. Instead, these foolish young people maintained their innocence! (They were probably holding on to some ancient and apparently outdated idea in our American philosophy of being "innocent until proven guilty.")

This incident was immediately portrayed as "Evil Big Rich White Guys" taking advantage of the "Poor Good Little

Black Girl." As we know, the Big One is always evil and the Little Guy is always the innocent victim. The *Moral Outrage!* toward the La Crosse players immediately exploded all over this story.

Judge Nifong used this case as a means of getting himself re-elected as a prosecutor who was "tough on crime." Jesse Jackson immediately stepped in to express his *Moral Outrage!* that these three guys could do such a thing to this innocent black woman. He even offered this poor victim a college scholarship. A spate of Duke Professors took out an ad in the paper excoriating these guys for their actions. And the La Crosse coach lost his job over the unsavory incident.

Jump ahead to the following April.

The La Crosse players were declared innocent. The stripper made it all up.

Around the same time all of this unfolded, the Rutgers women's basketball team finished a wonderful season for which they ought to be proud. In commenting on their season, Don Imus made a tasteless comment about the appearance of these girls.

Moral Outrage! hit the fan!

The tape of Imus' comment was replayed ad nauseum across the country. All commentators began every observation with the requisite denouncement, "What Imus said was despicable and disgusting," and then went on to make whatever moral observations they considered appropriate.[30]

---

[30] Talk radio hosts in particular found this incident "CHILLING" because they realized they could easily be out of work if one their own tasteless comments were blown out of proportion. It was these same

Being found guilty in the court of Moral Opinion, Imus was then required to go before a Judge and beg for forgiveness. He chose the Rev. Al "Racebaiter" Sharpton as his qualified place of absolution. He went on The Good Reverend's radio program. Apparently he did not appear sincere enough in his apology to the Good Judge Reverend Sharpton, because the sentence handed down was, "Forgiveness denied!"

And Don Imus took early retirement.

Notice the CONSISTENT APPLICATION of *Moral Outrage!* as each incident unfolded.

Also notice the INCONSISTENT APPLICATION of Moral Standards. Don Imus lost his job over a tasteless comment about the appearance of the basketball team.
Jesse Jackson and the Duke Professors got off scot-free in using the Court of Public Opinion to falsely accuse, try, and convict the La Crosse players of gang rape. Apparently there is no apology needed from any of them.

Bill Clinton, the D.C. Madam, and the Wal-Mart CEO

The "D.C. Madam", Deborah Jean Palfrey, released her 46-pound clientele list to *ABC*.

Moral Outrage! exploded!

Republican Randal Tobias immediately joined Imus in adding to the unemployment numbers in America, because he was named by The Madam.

---

media types who were making their living off of the story. They are the ones who got Imus fired by keeping it before the public and replaying it the requisite 32,000,000 times.

What's the big deal? We already determined with Bill Clinton that what happens in people's PERSONAL LIVES is to be completely divorced from their public responsibilities. So the whole sex thing is not worth discussing.

So then, let's consider *The Income Ratio Analysis* for determining morality. This analysis measures morality based upon the income ratio of the top individual in an organization with the bottom rung employee.

As reported on *NPR*, the measure of morality of the CEO of Wal-Mart has declined dramatically over the years. A number of years back the CEO earned about twelve times the salary of an hourly wage earner. That was declared acceptable. Presently, the CEO of Wal-Mart earns about 400 times the salary of the hourly wage earner. This is deemed terribly wrong. Grossly immoral!

By applying this same standard, Deborah Jean may in fact be THE MOST MORAL EMPLOYER in America. She made an even split with her workers. On a 300 dollar charge, she kept 150 dollars and so did her "employee." Measuring morality by the *Income Ratio Analysis* method, I say the D. C. Madam should be applauded for her high ethical standards! In fact, she is 400 times more moral than the CEO of Wal-Mart.[31]

(If we consistently apply this method of measuring morality, we are driven to the surprising and sad realization that the most immoral employers in history were the organ grinders. The little monkey did the tricks that delighted the crowds

---

[31] Before this book went to press, Deborah Jean committed suicide for the pressure brought to bear upon her supposedly immoral behavior. That was truly sad. It is unfortunate that no one applied the *Income Ration Analysis* to her business plan to relieve her of such public condemnation.

and then held the cup for the donations. People weren't contributing because of an old guy turning a crank. They gave because the monkey was cute. And how much did the organ grinder give the little monkey? BUPKIS! Infinitely immoral.)

****Newsflash!****

Paris is going to jail!

Having previously been pulled over for drunken driving and then more recently driving on a suspended license, she was sentenced to 45 days of jail time when she was caught speeding at night without headlights.

This raises serious and important questions.

Who will Britney party with now?

Will Paris co-author a book with Martha Stewart?

Does Geraldo think Paris is exploiting the Paparazzi in this particular stunt?

Chapter 26

Smart Wars

It was inevitable.

After the *TIME* article on "Evolution" tried to outsmart the *Newsweek* article on the "Death of Pluto," *Newsweek* came back swinging like a punch-drunk Rocky Balboa. The only question was, "What new burning topic would they choose to re-establish themselves as *The Smartest Weekly News Magazine in the World?*"

Perhaps they would tackle The Global Warming issue. Oh, wait! That goes contrary to the Global Cooling Scare they printed just three decades ago. Flip-flopping on this issue could make them look like John Kerry, and looking like *him* is anything but smart.

Or, they could report on the progress of embryonic stem cell research. Oh wait! It is only adult stem cell research that has produced any real results thus far.

Better yet! They could report on how the unbridled

invasion of the Zebra Mussels in the ballast water of the big tankers is destroying the Great Lakes. Oh, wait! That could be used by someone to draw parallels to the negative impact of unbridled illegal immigration. Even the politicians aren't stupid enough to tackle that problem. And nothing could look more stupid than being dumber than the politicians.

So, the smart people at *Newsweek* did the only thing they could do. They tried to outsmart *TIME* by creating their own *cutting edge* "Evolution" story. Now, they didn't say outright that is what they were doing, but a quick comparison will show that this is indeed what was at hand.

First, they kept hinting that the *TIME* article was so "Old School" in its understanding they might as well have proclaimed the world is flat. Whereas the *TIME* article showed a very simple and direct family tree from chimp to man, the *Newsweek* article contended that we now know that depiction is far too simplistic. The NEW understanding is that there was no direct descent between chimp and man. The progress is better characterized by "fits and starts." A graphic with about 651 different skulls and lines crisscrossing back and forth was included to show all the intricate relationships. This was a very clever ploy, because nobody could possibly understand this chart or argue that it is incorrect. In fact, this chart is so complex it allowed *Newsweek* to substitute the use of *Graphics Overload* in place of *Statistical Overload*.

Newsweek then takes the issue of brain capacity to a new level with the discipline of *Paleoneurology*. Everyone has been taught that larger brain capacity was a linear progression, resulting in advanced intelligence, leading to tools (like chopsticks) and thoughts (like morals). But now, calculating brain capacity is too simple. It is so "Last

Mesozoic." Catch this insightful comment: "Besides revealing the size of a brain, Paleoneurology examines impressions of surface features that the brain leaves on the *inside of the skull*." [#] (Emphasis by my wife, Laurie) Thus, by examining the bumps and ridges on the inside of the skull, paleoneurologists are able to tell our *past*.

This makes perfect sense because Paleoneurology is simply the "flip side" of *Phrenology*. Phrenology is the highly respected study of the bumps and ridges on the *outside of the skull*. From these contours, those accomplished in the art of phrenology are able to tell one's *future*. Statistically speaking, Phrenology is considered significantly more accurate than palm reading and only slightly less accurate than a crystal ball.[32]

Clarifying these two disciplines looks like this:

Paleoneurology: bumps *inside* of skull reveal the past

Phrenology: bumps *outside* of skull reveal the future

After utilizing Paleoneurology and creating Graphics Overload, the *smart people* at *Newsweek* then tweaked the *Linguistic Charade* of *dark matter*. This is the aforementioned standardized method of admitting one's ignorance while attempting to maintain one's intelligence.[33] Close observation reveals the term *dark matter* was never once used in the article. Notice this line, however:

> But now, as ancient DNA and *gray matter* give up their secrets, they are adding life to the age-old quest to understand where humankind came from

---

[32] Google "Phrenology" for the many sights referencing this discipline.
[33] See chapter on **Symbiosis**.

and how we got here. * (Emphasis by my goldfish)

Gray matter is lighter than *dark matter*. It is lighter because it is closer to the light of truth. Therefore, it is to be perceived as *smarter*.

Just in case we "slide rule dumb" readers did not catch on to the fact that the *Newsweek* people are SO MUCH SMARTER than the *TIME* journalists, they also sought to impress upon us that they are edgy, avante garde, and always willing to "push the envelope." Whereas *TIME* offered TITILLATING DISCUSSION on the inter-species sexual expressions of pre-human primates, *Newsweek* took it to a whole new level by offering FULL FRONTAL NUDITY! A male and female Homo erectus are pictured walking side by side.[34] WARNING! If you decide to check this out for yourself, this couple is uglier than the nudity of the fifty-year-old men shown in the movie *Wild Hogs*.[35]

Worthy of note is the fact that shades of The Pluto Problem surfaced in this VERY INTELLIGENT AND UNBIASED observation made by Sharon Begley and her co-author Mary Carmichael:

> The realization that early humans were the hunted and not hunters has upended traditional ideas about what it takes for a species to thrive. For decades the reigning view had been that hunting prowess and the ability to vanquish competitors was the key to our ancestors' evolutionary success (*an idea fostered, critics now say, by male domination*

---

[34] Close observation of the graphic leaves one wondering why they are called Homo erectus. But I guess the *smart people* at *Newsweek* know better than me.

[35] Can anyone please explain to me what movie ever enhanced its quality by showing the bare butt of a 50-year-old man?

of anthropology during most of the 20th century). (Emphasis by Gary, SpongeBob's pet snail)

One is immediately struck by the reality that this is The Pluto Problem all over. Everything these *smart people scientists* have been telling us dummies for years about our hunting capacity is now considered in error. But we can certainly trust them now that they finally have it right.

Clearly, the reason for the error was the men's fault. If women were involved they would have GOTTEN IT RIGHT THE FIRST TIME! We could have avoided these many years of monkeying around with this Great White Hunter nonsense. It was our nurturing, not our hunting, that saved us. Stupid Men!

****Newsflash!****

Good News!

Britney is out of rehab!

It worked!

We know this immediately, because she is going back to the recording studio for the first time in four years.

She has made so much progress! In the pictures we are seeing now, when she is unclothed, she has the sense to cover herself with her hands.

You go girl! You have become the model of integrity to your children!

Now I can return to writing this magnificent and insightful work knowing You, Paris, the Press and our Beloved Highly Moral Politicians will supply us all with an endless stream of *Moral Outrage!* and TOTALLY INCONSISTENT MORAL PRONOUNCEMENTS.

Chapter 27

Transcription Error

While waiting for Britney to exit rehab, I e-mailed *TIME* and *Newsweek* seeking permission to reprint the necessary parts of their articles for publication in this work. *TIME* e-mailed back with a nice form letter that totally missed the point of my request. I can forgive them, because they were probably working really hard on their list of the *100 Most Influential People in the World*. The list was published the month after my request was made to them.

Oh, the wonderful and exacting job they did! George Bush was left off, but Raul Castro made it on. Let's see! The president of the most powerful country in the world is less influential than the temporary substitute dictator of a small Communist Caribbean Island. No bias in that list at all! Isn't it good to know we can trust these HIGHLY MORAL NEWS PEOPLE to not let their personal perspectives color the hard news they create…Um, I mean report!

Chapter 28

mRNA

Morality, the component that makes us human, had to come from somewhere. Only two possible sources exist for the morality we embrace. It either became a part of our experience by random chemical processes or was programmed into our psyches by an outside moral entity, a Moral Designer of sorts.

Whichever the actual source of our moral perception, unique implications present themselves for each source.

'Splainin' the Helix

Chapter 29
Two Roads

Two roads diverged in a yellow wood,
And sorry I could not travel both
And be one traveler, long I stood
And looked down one as far as I could
To where it bent in the undergrowth.

Then took the other, as just as fair,
And having perhaps the better claim,
Because it was grassy and wanted wear;
Though as for that the passing there
Had worn them really about the same.

And both that morning equally lay
In leaves no step had trodden black.
Oh, I kept the first for another day!
Yet knowing how way leads on to way,
I doubted if I should ever come back.

I shall be telling this with a sigh
Somewhere ages and ages hence:
Two roads diverged in a wood, and I--

> I took the one less traveled by,
> And that has made all the difference.
> -- Robert Frost, 1916

This well-known poem was more than required reading in Mrs. Rich's eighth-grade English Class. We had to memorize it and then recite it while standing in front of the class.[36] Though famous for this poem, Robert Frost was neither the first nor the last to speak of the "Road Less Traveled."

Nineteen centuries before Frost, a man some would consider the World's Greatest Moral Teacher, spoke of a wide way that many traveled. Their end was destruction. He challenged his followers to pursue the straight and narrow path. Few would take this Road Less Traveled, even though it was only on this path that good reward could be found.

A century after Frost, *smart people scientists* are also making similar assertions. *Newsweek* states:

> ...humankind's roots are sunk deep in the East African savanna. There, the last creature ancestral to humans as well as chimps—our closest living cous-

---

[36] Seeking to add visual enhancement to this assignment, David Johnson walked around the room while reciting the poem. He became the "Walker" in the poem who traveled his own road. He timed his gait to arrive back at his desk just as the poem finished. He was a few steps behind so he had to run to his desk as he finished. This sent ripples of laughter throughout the class. His technique worked, because his is the only recitation I can remember. Who knew that just a few months later I would be campaigning for Hubert Humphrey with David Johnson? Our own roads diverged shortly after that. David became part of the drug-using Hippies and I joined the Bowling Club. Last I heard he was having a successful career within the Chicago Democrat political machine. When we meet at a school reunion one day, I am going to ask him if he knows that The Picasso is a caricature of Chicago's most famous Democrat, Mayor Richard J. Daley.

ins—lived, standing at a fork in the family tree as momentous as it is contentious.

The apes that stayed in the forests hardly changed; they are the ancestors of today's chimps. Those that ventured into newly formed habitat of dry grasslands had taken the first steps toward becoming human.

These pilgrims were strikingly few... The best estimate: 2,000 men. Assuming an equal number of women, only 4,000 brave souls ventured forth from Africa. We are their descendants."[37] (Aren't you glad I didn't have to do the whole "Emphasis" thing?)

Whether The Moralist or The Scientist, each contends that there is a Road Less Traveled and taking this road does indeed make all the difference.

Because each road makes such a significant difference, this

---

[37] *Newsweek*, March 19, 2007, pp. 54, 58. See page 57 for a map that plots out the routes of these ancestors. The explanatory paragraph is labeled "Roads Traveled." If this assessment of our heritage is true, then it provides some interesting social ramifications. That means WE ARE ALL AFRICAN AMERICANS. George Washington was our first African American President. Sorry, Barack! Jackie Robinson was not the first African American to play Major League Baseball. Enrollment "quotas" at colleges based on ethnicity make no sense since we are ALL African Americans. It also means the guys at Augusta who wouldn't allow African Americans to golf there were paying BIG DUES to a Country Club from which they technically should have excluded themselves. And the KKK meetings known to be full of sheets also had a whole lot of people who were themselves, umm, "full of sheets."

difference needs some "'Splainin'."[38] Before seeking an explanation of these significant differences, we ought to take note that an unusual aspect of these relatively few 2,000 males who took the Road Less Traveled is enlarged ear canals. Scientists believe these enlargements were from constantly sticking their fingers in their ears. This was necessary to drown out the sound of their female traveling partners. After all, these guys were traveling in totally unsettled territory. They had not yet adapted to the new habitat, and at times their traveling companions were less than happy.

Following is an example of a conversation scientists believe induced the "fingers in the ears" behaviors:

Female: Hey, we are going the wrong way! Everybody else is going the other way.

Male: I know they are going that way, but I am taking the Road Less Traveled.

Female: We're lost, aren't we? Why don't you just admit it and stop for directions?

Male: We are on the Road Less Traveled. There is no one

---

[38] The concept of 'Splainin' these differences must be credited to Donald Johanson and Tom Gray. Folklore tells the story that the night they discovered some bones was spent in drinking, dancing and singing. Supposedly they named their world famous Ethiopian "Lucy" after the Beatles song *Lucy in the Sky with Diamonds*. This is not correct. Contemplating the significance of their find, they concluded they found "The Missing Link." They proclaimed, "This explains everything!" They then named the skull Lucy based on the *I Love Lucy* TV show when Ricky Ricardo would demand, "Lucy! You got some 'splainin' to do."

here to give us directions.

Female: Okay, Mr. Boy Scout! I need a potty stop. Find a place that's cleaner than that Pete's Place you took me to last time. The restroom was so dirty I couldn't even use it.

Male: You can go in the weeds.

Female: Go in the weeds? What if someone sees me?

Male: We are on the Road Less Traveled. There is no one else here to see you.

Female: I am tired of traveling. You better find us some food and a place to sleep for the night.

Male: We can sleep in this soft grassy area and eat these seeds.

Female: Sleep in the grass? Are you crazy? I need a tree, or I won't get a wink of sleep. And eat seeds? You know I only eat bananas! You are so inconsiderate! I am beginning to think Mother was right. She said you were a drifter who would never settle down. And look at where we are now. Out in the middle of nowhere on your Road Less Traveled, sleeping on the ground, eating seeds, doing our personal business in the bushes and…Hey! Why do you have your fingers in your ears?

Chapter 30

Rational Road

While EVERYONE embraces a moral perspective, people travel two distinct roads to arrive at the destination from which they explain the source of their moral perspective.

Rational Road is the path chosen by those who contend that everything in the living world can be explained through the random process of continuous minor changes in the DNA pattern of living organisms. Those changes that offered benefit to the organism were favored by the environment and allowed to reproduce. Those changes which were not helpful to the organism were lost. This process is known as Natural Selection.

The distinctive elements of the Rational Road perspective are that ONLY "CHEMISTRY AND MATTER" are involved through the RANDOM process of Natural Selection over a great length of TIME. This slow advance of change in living organisms is called "Evolution." It is considered a powerful and creative force, able to make whatever it chose or needed to accomplish its own ends. As described in *Newsweek*, "Evolution played Mr. Potato Head, putting different combinations of features on ancient hominids then letting them vanish until a later species evolved them."[39]

Rational Road travelers contend that, beginning only with "Chemistry and Matter," the powerful force of Evolution developed a self-replicating system of information storage and transfer (DNA/RNA). This self-replicating system eventually BEGAN TO PERCEIVE ITS ENVIRONMENT by highly complex systems of touch, sight, hearing, taste, and smell. With increased complexity, the organism not only perceived its environment but began to respond to it via INSTINCT. Further development brought the ability to think and contemplate its own existence and the possibility of life's PURPOSE AND MEANING. Eventually a CONSCIENCE was invented by random processes, and ideas of RIGHT AND WRONG influenced the organism's interaction with its environment. The critical thought in all of this is that at no point did the process create anything other than a more complex arrangement of "Chemistry and Matter." The entire study of base pair changes (so thoroughly discussed in the Dueling Smarty Pants articles of *TIME* and *Newsweek*) is merely a mapping of one chemical arrangement as compared to another.

In this context, the origin of morals is explained as something preferred by Evolution. Perhaps by working together, rather than competing with each other, the population insured its own survival.

Imagine two orangutans beating each other with clubs for

---

[39] *Newsweek*, March 19, 2007, p. 55. I'll believe Evolution was playing Mr. Potato Head when they discover a fossil with a mustache growing out of its ear or an eye growing where its mouth should be. This is the stuff everyone does when playing Mr. Potato Head.

the sake of mating with the attractive redhead in the next tree. A saber tooth tiger comes along and figures he's got dinner and dessert in one easy serving. But because of some mutation in these particular orangs, they turn to beating the tiger instead of each other. They tan his hide and make a fur coat for the redhead. She is so impressed by the coat she begins dating both of them rather than choosing between them. *Voila!* Morality is born and everybody is happy. The attractive redhead has two strong males to protect her young, and the males each figure the other guy is the father so they can have totally irresponsible sex which (we know from the TV show *Men in Trees*) is the single most important thought on any male's mind.

This scenario changes if the attractive redhead also mates with an 800 lb, 92-year-old silverback gorilla who owns about a billion banana trees. When he kicks off and she stands to inherit the banana plantation, every rhesus, chimp, and baboon who ever met the redhead is going to claim to be the father of the baby.

The Evolution of Calculators as Pick-Up Tools

Slide Rule Calculator – The "Size Matters" Method

Buster never calculated the effect of a "pocket-size" slide rule on his social agenda.

Chapter 31

Lincoln Logs and Erector sets

As a young boy, I had Lincoln Logs. Lincoln Logs provide for hours of fun and creativity. I had enough logs to make a tower or a two-room house, and then I pretty much ran out of pieces. When my friends brought over their sets, we could work together and make bigger, more complex structures of the forts and castles varieties. Like Evolution playing Mr. Potato Head, we could make something, get rid of it, and make it again later. Theoretically, we could combine 50 boxes of Lincoln Logs and make a model of the World Trade Center. But no matter how complex our arrangements of Lincoln Logs became, they never morphed into an Erector Set. Whenever we took the Lincoln Logs apart we still only had wooden pieces. Nothing changed.

Being that I was raised on such "low tech" toys as Lincoln Logs, my cognitive skills were perhaps stunted by my experience and observations. The fact that my Lincoln Logs never became an Erector Set causes me to conclude that non-moral matter, regardless of the complexity of its arrangement, does not produce moral matter.[40] But if non-moral matter cannot produce moral matter, then how can all men behave according to moral principles?

First, *morality as a concept* of "good and evil" or "right and wrong" disappears as fast as the waiter in a cheap restaurant after being tipped. Regardless of how hard you search, non-moral matter is all you find on Rational Road.

And *morality as a standard* by which we live becomes as ethereal as the credibility of your average politician. "Good and evil" or "right and wrong" are no longer rational options. Lincoln Logs do not become Erector Sets and "Chemistry and Matter" do not spontaneously generate morality.

So, what is this energizing "moral compass" which we respect in those who have it, we denounce in those who don't, and we seek to live by in our own lives? Our moral compass is simply another "sense" selected for by Evolution. It is no more "moral" than our ability to see the color blue, or hear the old pop song "Blue Moon," or to smell the bleu cheese in our favorite salad dressing.

Our "moral compass" can have no real connection to ideas of "good and evil" or "right and wrong," because these are

---

[40] While speaking with some young people, I learned they all knew what Lincoln Logs were but weren't familiar with an Erector Set. That got me thinking. Perhaps as a marketing strategy, more companies should tie a president's name to their products in order to enhance name recognition. Thus, we could have the Clinton Erector Set (no explanation needed there!). Or, how about the Bush "Clue"less board game? The object of the game would be to find out who would be the next terrorists attacking our country, what population center would they attack, and where did they hide the WMD that everybody knew were in Iraq?

not rationally possible ideas in a universe made solely of "Chemistry and Matter."

Our "compass" is only an acquired ability to align ourselves with our surroundings according to behaviors that have been selected as advantageous for reproductive purposes within our environment. It is similar to the "safety instinct" for self-preservation that was selected for shortly after we branched off from the chimps and headed down the Road Less Traveled.

Imagine that because everyone loves Lucy, another significant archeological find had generally been overlooked. What if, in the African Savanna they discovered a dried river bed that revealed some interesting footprints and one complete skeleton. Two sets of chimp footprints are seen walking together. Coming directly at them from about thirty feet away are a distinct set of brontosaurus prints. One set of chimp prints steps aside for the dinosaur while the other had no instinct to do so. That set of footprints ends as a crushed skeleton within the advancing dinosaur print. So, scientists aptly call that skeleton *Africanus flatasapancakous*. The fact that one individual developed an instinct for self-preservation and stepped aside for the dinosaur didn't make him moral. It only allowed him to reproduce. And so, too, for whatever behaviors we evolved that we claim are moral. They are not truly moral. They are simply instincts that allow us to reproduce.

Chapter 32

Global Warming

Global warming is the *calamity du jour* among the modern illuminati who are able to predict the future for Mother Earth. They tell us all of the problems we can expect if we do not address this issue NOW! Al Gore authoritatively declared Global Warming THE MORAL IMPERATIVE of our day. Way to go, Al! But before I give up my affinity for golf so I can afford to buy some of your "Carbon Offsets", could someone please answer a few questions?

I accept that it is getting warmer, but I live in northern Minnesota and I will take ALL THE WARMTH I CAN GET! My desire to see warmer winters leads me to ask, "What's wrong with warmer weather?" How do we know the temperature we are at right now is the ideal temperature? How does one say our present thermal average is the ideal temperature of the earth? How does anyone know that?

Also, where does someone get the authority to say this is THE MORAL IMPERATIVE of our day? How is Al Gore qualified to assume this responsibility? Who made him the arbiter of such issues as abortion, immigration, gun control, and the pizza wars between Chicago and New York so that he is able to say Global Warming is THE MAIN ISSUE?

And if, as the *smart people* at *Newsweek* contend, climate changes accelerate the rate of Evolution, why do we not embrace the concept of GLOBAL WARMING AS A GOOD THING? Can we not expect the global warming we are introducing to induce some evolutionary changes? Historically, the climate has changed before. What is wrong with it changing now? Sure, some polar bears may die, but nature has killed off thousands of species on its own. We don't fault nature as immoral. And if we can increase the rate of Evolution so we can observe it, think of how much more we will be able to learn about life as we know it.

These questions about Global Warming raise parallel questions on the topic of morality. For starters, LET'S TALK ABOUT SEX—since it appears to be the issue that pervades our minds at all times as Americans. How do we know the ideal standard of sexual morality? (How hot should it be?) Is more sexual restraint to be preferred to more sexual expression? How do we know? Where do we start? How do we know our starting point on sexual morality (whatever and wherever that would be) is the right starting point? Who gets to set the standard? Who is the "Al Gore" of sexual morality who will declare for us THE SEXUAL MORAL IMPERATIVE of our day? How is this person qualified to make that determination?

How do we even know establishing sexual moral standards is a good thing? So what if sexual predators surf the web looking for innocent children who might make easy prey? So what if a raid in Minneapolis revealed a sex slave trade? Who says these are bad? Perhaps these sexual behaviors are the normal result of variation through Evolution and a necessary component to our further development. Why deem it immoral? Maybe it is simply one of those many Mr. Potato Head features that Evolution threw together for the fun of it.

Who exactly is capable of establishing morals and to what do they anchor them?

It appears to this Lincoln Log building, mentally stunted Baby Boomer that trying to establish moral standards on Rational Road is like asking Captain Barbossa's monkey to tie up The Black Pearl in the middle of the ocean during the Maelstrom sent by Calypso. Is he capable of tying it up in the first place? Even if he is capable of tying it up, to what would he tie it?

****Newsflash!****

While Paris' lawyers are trying to keep her out of jail and Britney is in the recording studio (and not speaking to her Mother according to the current scuttlebutt), the IMPORTANT NEWS about celebrity life meltdowns is rather thin.

Thankfully, Rosie was in another fight.

This time she picked it with another guest on *The View*, Elizabeth Hasselbeck. Elizabeth is less well known, smaller, and less powerful than The Trumpmeister. No one is saying that's why Rosie chose her for this battle, but it did get wild enough that it made the every-hour-replay on the 24-hour news shows.

Chapter 33
Enigmas of Virtual Morality

Being a slide rule and Lincoln Log, low-tech kind of guy, I was amazed to be exposed to the world of Virtual Reality at Downtown Disney in Orlando, Florida. My daughter and I designed and then rode a roller coaster without ever going anywhere. The simulated ride GAVE THE IMPRESSION that we were in reality riding a roller coaster. There really was no roller coaster, and we weren't really riding any rails through loops and dips and over big hills. That is the nature of Virtual Reality. You actually feel like you are experiencing a real world, but in reality the things to which you are responding do not exist.

Likewise is the morality on Rational Road. It can only be considered **Virtual Morality**. It appears and feels like moral reality, but in truth there can be no morality. Everything is "Chemistry and Matter." Nothing more. Nothing less. Morality is something we presume exists, but just presuming it, doesn't make it any more "actual reality" than the "virtual reality" roller coaster.

Rational Road ultimately leads to a morality dead end best described in the immortal words of the Chicago Biker, "Iddohmadduh." Consider the following test cases.

Test Case #1: Wal-Mart

Some guy named Mike is a major contributor to the Wake-up Wal-Mart website. Recently interviewed on TV, HE insisted that Wal-Mart needed to provide health insurance to its employees. When asked why that was necessary, in light of the fact that there was no law requiring such a provision, HE simply stated that Wal-Mart had a "moral imperative" to provide health care. Now, that sounds emotionally satisfying, but on Rational Road, how does one make that assertion? Is this guy capable of making the pronouncement that Wal-Mart has a moral imperative concerning health insurance? (Is the monkey capable of tying up the ship?) How does he know what to define as moral? What is his moral mooring? (Where does the monkey tie up the ship in the middle of the ocean?) Who is HE to say Wal-Mart must act according to this moral injunction as HE sees it? Who is HE to even insist that morality is a reality? If we are only "Chemistry and Matter," nothing more and nothing less, then where does morality enter into this discussion? It is perceived, but it is not real, because "Chemistry and Matter" are not moral. Iddohmadduh.

Test Case #2: Gitmo

We have been told for years that we are wrongly treating the prisoners at Gitmo. Yet, they have survived quite well because of the health care they receive in prison.[41] This care is free and far superior to anything Wal-Mart employees receive. If providing quality medical care is the foundation for morality, then the Gitmo detainees are being treated magnificently! But the real question is, "Who are these people to say that how prisoners are being treated at Gitmo is morally wrong?" Based on what standard? How do we know that standard is accurate? How do we know these people are qualified to establish and interpret the standard? Finally, how de we know the morality we seek to apply to Gitmo is real? If we are only "Chemistry and Matter", there can be no real morality. There is only Virtual Morality. It feels like something real. We respond to it as if it's actually there, but it doesn't exist other than as a conceptual world we have created for our own purposes.

One could list dozens of test cases that all lead to the same enigmatic questions.

The first enigma of Rational Road Virtual Morality is that we so readily believe morality is "actual." How could non-moral matter produce morality?

The second enigma of Rational Road Virtual Morality is that to embrace morality we must act immorally. Richard Dawkins, in discussing the issue of morality as a component of our evolutionary makeup, stated that we can all agree that murder is wrong. [#] THIS FEELS RIGHT and therefore we may embrace it as moral.

But the *smart people newscasters* tell us that a recent poll

---

[41] On the day I am writing this, the family of PFC. Joseph J. Anzack announced that their son's body was located and identified as one of three missing soldiers taken in an ambush in Iraq. He didn't even last two weeks in their health care. I AM CERTAIN that those who decry our immorality at Gitmo, will likewise be categorically denouncing those who killed this soldier.

indicates 26% of Muslim AMERICAN males ages 18-30 believe murder by suicide bombing is a morally acceptable practice to advance the cause of Islam. Clearly, these young Muslims tie their boats to a different mooring in the middle of the Virtual Morality Ocean than Richard Dawkins. Who is to say which mooring is correct?

If society agrees that murder is wrong, that still doesn't mean it is wrong. There was a time when a great part of society believed slavery was acceptable. That did not make it morally right. Society is only comprised of people, and who is to say any of these people are capable of defining morality any more than Al Gore? Their perceptions of moral behavior amount to nothing more than their opinions.

Besides, when society affirms that murder is wrong, they then OUTLAW THE ACT in order to get people to conform to their opinion. If people do not conform, they are locked up and in some places are given the death penalty. This is a form of slavery. People are being enslaved to act according to someone else's whims and opinions and against their own desires and beliefs. We claim slavery is wrong and yet we enslave others to our opinions. We justify this logical inconsistency by insisting it is the appropriate thing to do, because it is the "moral" thing to do. But according to whose morality? If Muslims were in the majority, would it make it morally right for them to legalize the killing of non-Muslims? A majority opinion is a flighty and unstable means of defining moral standards. Anyone with a charismatic persona could persuade others to his beliefs, and moral standards would change. This is acceptable on Rational Road as a means of establishing societal norms, but it cannot be claimed as "Actual Morality." It is only "Virtual Morality." It is a world which we create for our own purposes that does not exist in and of itself.

A third enigma totally boggles my low-tech brain. Why all the anger on Rational Road? Richard Dawkins clearly and accurately admitted that Evolution neither owes nor offers us any real purpose in life, because we are simply the result of a series of totally random and impersonal chemical reactions.*

While commenting on the death of Rev. Jerry Falwell, Christopher Hitchens (known atheist spokesman and famed traveler of Rational Road) expressed nothing but anger and vitriol for the deceased whom HE JUDGED to have foisted SUCH GREAT EVIL upon his fellow man. His contention was that Falwell did nothing more than start a big business that he then left to his family with his passing. The Good Reverend's business was predicated upon FOISTING ANCIENT MYTHS on people and CONDEMNING TO HELL all those who did not agree with him. Mr. Hitchens vehemently expressed that he thought it was unfortunate there was no Hell for The Reverend to rot in for ALL THE EVIL that he had done!

If we are nothing more than "Chemistry and Matter," how can Mr. Hitchens believe evil exists (and was perpetrated by Falwell)? Evil is a moral concept, and morality cannot exist in a purely "Chemistry and Matter" environment.

And how does Angry Christopher decide that The Reverend Business Mogul behaved in an evil manner? How does he measure what is evil and what is not? How and when did he become the authority on that which is evil? Certainly he is entitled to hold an opinion contrary to the Deceased Preacher. But does a difference of opinion require such animosity? C'mon, Chris! Why such anger over something that is as ultimately meaningless as life as you know it?

Even if Mr. Falwell built a business on propagating myth,

why is that evil? Is it not possible that some people need the myth as much as Good Christopher needs to reject any concept of a god? Because they "buy the myth" (by believing in a god) they buy the stuff which supports his business (by attending his school, contributing to his church, buying his books). Should it not be interpreted on Rational Road that the ready belief and propagation of THE MYTH is simply the result of evolutionary variation? Why define any of it as evil? We are "Chemistry and Matter." Concepts of good and evil, which are morally based concepts, are merely our own contrivances and amount to nothing more than Virtual Morality. Surely the *smart people* of the world, like Christopher Hitchens, can rationally conclude these variations do not matter. So why do they get their undies in a bundle over this stuff?

And furthermore, why do Rational Road wayfarers get so upset at the hint of teaching THE MYTH along side of Evolution in public schools? Why don't they simply teach the idea that a number of people believe THE MYTH and that this serves some unknown evolutionary purpose? Certainly the desire for truth on the subject would require such honesty. Besides, the quest for TRUTH can't be all that important on Rational Road. Our existence serves no ultimate purpose for which pursuing the truth on this topic would matter. We are only "Chemistry and Matter." Nothing more. Idohmadduh.

Speaking of being upset over nothing, what is going wrong with the world of great apes? In the news this week were reports of a 500-pound orangutan escaping from a Japanese zoo and a 400-pound gorilla escaping from a German zoo.[+] What have these guys got to be so worked up about that they'd go on the lam? They have better health care plans than Wal-Mart employees.

****Newsflash!****

Rosie quit weeks early!

Not that it matters one iota for anyone's life, but the *smart people newscasters* are really gonna be in a pickle when it comes time to give us our weekly dose of celebrity insanity.

No Anna Nicole, Paris, Britney, or Rosie.

OH NO!

What will they do if they have to report REAL NEWS to us?

Chapter 34

Designer Drive

Different strokes for different folks.

Whatever blows your hair back.

There's more than one way to skin a cat.

Whatever floats your boat.

A penny saved is a penny earned.[42]

These folksy quips indicate there is more than one way to "Git 'er done." Though oodles of people travel **Rational Road** to stake their claim on a Moral Perspective, an equally large number travel **Designer Drive.**

The sojourners on Designer Drive perceive their Moral

---

[42] Frankly, this quote has nothing to do with the others, but it has been attributed to Ben Franklin and I thought quoting him would add some *gravitas* to my writing.

Compass to be the result of non-random processes. They contend that a presence/entity/force which is *something other than "Chemistry and Matter"* is at work in the universe. This Essence is a **Moral Designer** which impresses morality upon our human existence in some way. Whether it is karma, reincarnation, the God of the Bible, the good and evil spirits of tribal cultures, ancient myths, or any one of such countless perceptions, the salient distinction is that The Moral Designer is something *other* than "Chemistry and Matter."

The Designer Drive types embrace a concept that says the Moral Designer has both the Capability and Authority to establish Actual Morality. This is because He/She/It IS THE SOURCE of Morality. The *moral compass* that guides these travelers is the morality they believe has been authoritatively established by the Moral Designer.

Designer Drivers believe the Moral Designer is capable of defining right, wrong, and even moral imperatives in a way that Al Gore cannot do. The Moral Designer does not simply voice an opinion. The standards established by The Designer become the framework for Actual Morality.

The Designer is also capable in some fashion of enforcing the standards. This enforcement is not deemed immoral, because it is Actual Morality and not mere personal opinion being forced upon individuals with contrasting wills or desires. Christopher Hitchens would be glad to know that on Designer Drive the Moral Designer has prepared a punishment capable of bringing justice to Jerry Falwell for the evil he has done.[43]

---

[43] An ancient Hindu story describes how justice is meted out in an Actual Moral world. Professor Sadist taught Physics at India University. He enjoyed making monkeys squeal, so he used live monkeys for his

gravity demonstration. Since dying he has been repeatedly reincarnated as a termite outside of Pete's Place. He keeps getting eaten alive by monkeys with sticks until his dead termite body mass equals the body mass of the sum of monkeys used in his demonstrations. It has been calculated by the people at Orkin that his termite reincarnation cycle will end about the same time Congress fixes Social Security.

Chapter 35
Creek Jumping

I was a Pre-Pong college student. Pong, the first home video game, didn't make its debut until after I graduated. Nobody had PCs or laptops. No sitting around a dorm room playing *Halo III* by the hour. No DVD movies to rent. No Internet.[44] That meant when we were broke (which was always) and looking for something to do (which was most of the time because we didn't want to bother with studying physics), we had to come up with some inexpensive low-tech means of entertaining ourselves. So we developed the sport of Creek Jumping.

The rules to Creek Jumping were simple. We went to a creek running through the South Farms. These were the experimental grounds for the Ag School. We walked along the creek until someone got a hankering to jump across. When one guy jumped the creek, everyone else had to follow. The goal was to select a spot where hopefully some-

---

[44] Al Gore was still working out the bugs and had not gone public with it yet.

one would get wet. It sounds pretty stupid on paper, I know, but then we're guys and it doesn't take much to entertain our moronic brains. Just ask all those lady authors at *Newsweek*.

Regardless of whether one travels **Rational Road** or **Designer Drive**, everyone has to do some Creek Jumping in order to stake their moral claim. The creek we all jump could well be called *Mystery Creek*.

The Rational Roadies jump *Mystery Creek* at the point of believing that non-moral "Chemistry and Matter," random processes, and enough time produced a moral system. They say our Moral Compass came into existence through the evolutionary process. But there is no reality behind the morality. It is Virtual Morality. The existence of morality is a TOTAL MYSTERY since we are only "Chemistry and Matter." Tourists on Rational Road simply assume a Moral Framework because it seems right. But in reality, Iddohmadduh.

The Designer Drivers jump *Mystery Creek* at the point of believing there is something other than "Chemistry and Matter" that infused a Moral Compass into our existence. This compass gives direction through Actual Morality. How this Moral Designer experiences its own existence is the mystery they embrace. They simply accept the presence of a Moral Designer, because it seems right.

Regardless of which road we travel, as moral beings we all have to jump *Mystery Creek* at the point where we accept something about the source of our morality which we cannot prove.

The place where one jumps *Mystery Creek* is the place where one answers for one's self the age old question,

"Which came first: The Monkey or The Morals?" Rational Road proponents claim The Monkey came first, while those who travel Designer Drive claim The Morals were first. Having jumped the creek at the place of their choice, they then point at those of the alternative persuasion and say "they are all wet."[45]

---

[45] They also tend to think and speak of the others as the "Embodiment of Evil." This allows them to FEEL MORALLY SUPERIOR and reinforces in their minds that they have chosen the good, right, just, and true Road Less Traveled.

****Newsflash!****

Without Britney, Paris, Rosie, or Anna Nicole in the news I am really hard up to find material that will entertain you enough to keep you reading. So I have to sink to doing what the politicians do in order to keep themselves in the news. I have to voice my *Moral Outrage!*

When I began this project eight months ago, monkeys were nowhere to be seen. Last week I saw two reports of escaped great apes. I have also noticed that THREE DIFFERENT COMPANIES ARE CURRENTLY USING MONKEYS IN THEIR TV COMMERCIALS![46]

Just a coincidence? I think not!

All of this stuff broke onto our national scene sometime AFTER I wrote the *smart people* at *TIME* and *Newsweek* for permission to use their materials. I think they leaked my idea on this project, and it was obviously seen as so exceptional, that these companies got in on the whole MONKEY THEME before I came out with this project so they were in a good position to capitalize on the impact of my work.

I will be seeking legal counsel to have this INJUSTICE addressed!

---

[46] Taco Johns with "Whiplash," CDW with the technology chimp, and Capitol One with the monkey-on-your-back chimp. By the time we went to press, this list had grown even further, with a full length Hollywood movie seeking to cash in on the impact of this book!

Chapter 36

Comparative Anatomy

Instead of Creek Jumping, I probably should have been studying for Comparative Anatomy. In that class we studied the structures of various animals to observe their similarities and differences. The idea was to see how a fish flipper became a lizard leg which became a buffalo wing. In one particular lab, we moved from station to station looking at various animals. Each one seemed to me to have highly complex and fully developed appendages perfectly formed for that particular creature. We never really saw anything that was an actual transitional form but the *smart people professors* told us we could assume that these transitional forms existed at some time.[47] All of that was, of course, OLD SCHOOL evolutionary teaching. We now know Evolution is not linear but prefers playing Mr. Potato Head and changing parts all willy-nilly-like, so studying that stuff was about as valuable as learning Pluto was a planet.

---

[47] One of the stations in a particular lab had a Ratfish. We were told it was caught somewhere off the Left Coast. It was appropriately named because it had a face that resembled a rat. It was REALLY UGLY! I mean, REALLY UGLY! You could tell whenever someone arrived at the Ratfish station because they took one look and said, "UGGH!!"

Playing Comparative Anatomy with the Rational Road and Designer Drive perspectives may also be no more valuable than learning that Pluto was a planet, but we are bound to have as much fun as Evolution had playing Mr. Potato Head.

Rational Road

Our entire human existence can be rationally explained through an analysis of our "Chemistry and Matter."

We developed solely via impersonal random processes.

Life has no ultimate or universal purpose.

No Actual or Universal Morality exists.

Only Virtual Morality exists.

Morality *feels real* so we declare it to be so, even though there is no actual morality in play, nor can there be.

Our morality is sourced in our personal opinions and tastes.

Enforcing our morality upon others (laws) is not a truly moral position but a power issue, because no true morality exists.

Impressing our morality upon others is *nothing more than* impressing our opinions upon them.

Enforcing morality (laws) as a morally correct position is logically inconsistent in that it enslaves others to the opinions of our Virtual Morality against their own opinions and desires.

Ultimately, Rational Road endows the most persuasive or powerful Virtual Moralist to be the Supreme Ruler of a world that is meaningless.

Designer Drive

Our existence lived in the context of "Chemistry and Matter" is also infused with a moral compass by an outside Moral Designer.

We were intentionally developed by a Designer.

A Designer gives ultimate purpose to life.

Actual Universal Morality exists.

Real Morality exists.

Morality *feels real* because it is real and has been infused into our human consciousness and experience.

Our morality is sourced in our perception of The Morality.

Enforcing our morality upon others (laws) can be a moral position when it is an accurate expression of The Morality. Impressing our morality upon others *may* result in our impressing a real Morality upon them.

Enforcing morality (laws) as a morally correct position is logically consistent when the enforcement is in accord with The Real Morality as defined by The Moral Designer.

Ultimately, Designer Drive constrains the Real Moralist to be subservient to The Moral Designer in a world filled with meaning and purpose.

It would have been way cool in Comparative Anatomy to find something like the flying monkeys from the Wizard of Oz. If Evolution was really playing Mr. Potato Head for all those billions of years, throwing parts together and discarding them again like nobody's business, then surely it could have given us something really strange like flying monkeys. These critters would be readily selected for survival, because they would have all the advantages of flight, climbing trees, and using chopsticks.

****Newsflash!****

Paris is out of jail!!

After reportedly spending only three days in jail,[48] Paris has been released. We are being told it was a medical issue. Apparently she was sick of being out of the spotlight.

Technically she has been "re-assigned but not released." She has to stay home and wear an ankle bracelet. AMERICA WEEPS for her hardship! Her friends will be consoling her tonight with a party at her mansion.

The *Moral Outrage!* was immediate.

The Rational Road advocates have been pretending a great injustice has been done even though they know all that has happened is that someone powerful has used their power to strengthen their position. It is how the universe works. Power wins.

Tourists on Designer Drive believe a great injustice has been done as they perceive The Universal Moral Standard has been breeched.

Everyone feels good about their personal sense of *Moral Outrage!*

And the journalists can return to entertaining us!

---

[48] The authorities somehow say her stay was five days. The word on the street is they used a slide rule instead of a push button calculator to arrive at this number.

Chapter 37
Speed Bumps and Pot Holes

Regardless of which "Road Less Traveled" one chooses to take, whether Rational Road or Designer Drive, there will be speed bumps and pot holes to traverse. These things tend to make the trip a bit uncomfortable for the traveler.

Sojourners on Rational Road get bumped around by **Logical Inconsistencies.**

First is the inconsistency of making moral claims. If we are only "Chemistry and Matter" and the sum total of our humanity can be defined by the base pairs in our genetic makeup, then we are entirely non-moral beings, and our attempt to stake out our morality as something real is either an intentional or unintentional deception. Even the best thoughts we can muster are only a system of Virtual Morality built upon a non-existent foundation. They are opinions about right and wrong in a world where right and wrong do not exist. This is not to say that *feelings* of morality do not exist. The feelings are very real. It is the morality itself that does not exist. The feelings which seem so sure of their own "moral high ground" are nothing more than responses to the environment which have been selected for by Evolution.

Second, to claim one has the moral high ground means someone with an opposing perception is by definition missing the mark morally. Some form of power structure (laws, ostracism, excommunication, etc.) is then used to enforce conformity to the position of the stated moral high ground. But if morality does not exist in a solely "Chemistry and Matter" environment, then using power to enforce conformity to one's opinion is actually an enslavement and the antithesis of a moral action. Indeed, no use of laws or societal structure can ever be considered moral or just.

Third, there is no place to get a footing to begin building one's moral construct. Richard Dawkins claims we can all agree that murder is wrong. But bestow knighthood on Salman Rushdie (known for drawing a cartoon of Mohammed with a bomb in his turban) and you will find thousands of Muslims ready to commit murder to avenge his insult to Islam. They would not agree with Dawkins' starting point on morality. Who is to say they are wrong? This is nothing more than one man's opinion against another, with no objective standard by which to measure either of their opinions.

Those who choose Designer Drive as their route of choice are not straddled with the Logical Inconsistencies of their wayfaring counterparts. They believe the source of their moral construct comes from a Moral Designer. This Entity has the insight to establish what is truly right and wrong because He/She/It is the ground of all morality. If the Entity says, "Murder is wrong," then murder is wrong. It doesn't take long to see that there are deep **Practical Inconsistencies** with the proponents of such a view.

From Parents to Preachers to Politicians, people in places of authority who claim their morality is based upon some Supreme Morality, regularly violate that morality with barely a hint of remorse.

First, our country was founded upon Judeo-Christian values and morality.[49] With deep emotion we sang *God Bless America* after 9/11. Our public oaths end with "...so help me God." Historically, all politicians take their oath of office by placing their hand on a Bible.[50] And yet we have banned the Ten Commandments from public display and schools because someone might be influenced by them. Shouldn't that be a good thing if indeed these

---

[49] For those of you whose *Moral Outrage!* flares up at such an assertion, DON'T GET YOUR UNDIES IN A BUNDLE! I know you have push button calculators and believe you can prove **every person** who fought in the Revolutionary War and **all** the signers of the Declaration of Independence were Deists. I know you want to contend that there was no influence of Christianity in our early history. But if you could work with me for just a minute here, even though I only have a slide rule to figure this all out, perhaps you could acknowledge that all the biblical references engraved in the buildings and monuments around our Capital suggest that there was *at least the hint* of a Judeo-Christian world view among our forefathers.

[50] That is unless you are a Muslim from Minnesota. Here we elected a congressman who insisted he could take his oath on a Koran. He claims he can separate his Islamic background from his political life. I will believe that when he takes his oath on the traditional Bible. He also has been heard on Public Radio proclaiming his desire that other Muslims will run for office. If militant Muslims can justify running planes into buildings to achieve their end, does it at least seem possible that lying about the relationship between their faith and their political intentions is not a difficult thing for them to do? Should we perhaps exercise some extreme caution here? And if he truly separates his work from his faith, why does it matter if more Muslims get into office?

commandments came from The Moral Designer? We eliminated prayer from our schools.[51] Is my slide rule bent or is it incredibly inconsistent to ask The Moral Designer for protection immediately after 9/11, while throwing out every acknowledgement of the Designer's authority, presence, and rule from our public discourse?

Second, Designer Drive aficionados speak out of both sides of their mouths on morality issues. The same Politicians who claim a higher minimum wage is just and moral, want to keep the immigration floodgates open to provide cheap labor for jobs which "Americans won't do." How about raising the wages for those jobs so Americans will do them? Is it moral to underpay the migrant worker while insisting legal Americans receive a living wage? The same politicians who promised us the "most ethical congress ever" gloat over the imprisonment of Scooter Libby but won't go near Sandy Burger who stole and destroyed top secret documents. Barack Obama claims it is above his pay grade to know when life begins but he is more than willing to run on a platform that will define when abortions are legal.

How about parents who teach their sons not to cheat on tests but try to get them into the movie theater on a children's admission when they are 17 and have been shaving for three years?

Though logically consistent about the existence of morality, Designer Drive types are practically inconsistent in demon-

---

[51] Some people note an interesting statistical correlation between the removal of prayer from public schools and the increase in out-of-wedlock births. There is also a statistical correlation between poverty and single mothers. Of course it is only morally and politically correct to speak of the poverty issue. Any correlation between the prayer issue and out-of-wedlock births that cause poverty is simply not up for discussion.

strating that morality.

The fact of the matter is *everyone* is a miserable failure at living out the morality they claim to embrace. We seem incapable of missing those speed bumps and pot holes.

Here is where the monkeys and apes outdo us every time. They are always consistent. They work on the Alpha Wolf system. If you come across a gorilla in the jungle and he doesn't like you, he will rip your arms off. But he never makes a self-contradictory high falutin' sickeningly pious moral assertion about his behavior. He just rips your arms off because he feels like ripping your arms off. End of story.

****Newsflash!****

Paris is back in jail!

Seems some judge didn't buy her story, (or she didn't buy the right judge). Whatever happened, he yanked her "Get Out of Jail Free" card and sent her back to the slammer.

All those who were experiencing *Moral Outrage!* that she got special treatment just because she is part of the wealthy class, are now feeling that justice has been served.

Chapter 38

Tee Shirt Philosophy

Every so often I see a shirt in the mall on someone sporting a "six pack." Not the six-pack one earns from many hours in the gym. They are wearing the six-pack they drank while sitting on the couch watching football.

The shirt says,

> Everybody's gotta believe something...
> I believe I'll have another beer.

Whenever I see that shirt it reminds me that I am not one of the *smart people*, because it always makes me laugh.

Maybe my lack of intelligence is why I believe as I do.

When the choices for explaining our moral construct are either Rational Road or Designer Drive I always find myself meandering down Designer Drive. This is for four simple reasons.

First, it appears every bit as logical to embrace the idea of a Moral Designer as it is to embrace random chance as the progenitor of our morality. Just as wobbly stars suggest the presence of some massive black hole which is not seen, and similarities between animals supposedly suggest intermediate forms which have never been seen in the fossil record (even though Darwin himself said if his theory were correct they would be numerous), then the ubiquitous nature of our moral expression suggests the presence of a Moral Designer even if it is not seen.

Second, is Baby Camo. I recently traveled to New Orleans with a team of young people and adults to lend a hand with the massive amount of work that still needs to be done as a result of Hurricane Katrina. While there I saw BIG bugs and POOR people.

When a bug wandered into our bunkhouse, I demonstrated to my young friend Camo[52] that the best way to kill such a bug is with a flip flop. He wanted to kill it with a big work boot, because the boot was so heavy. This would seem correct, because these bugs are so big THEY NEED A GOOD WHACKING, and a work boot can provide considerable heftiness. But a work boot is cumbersome, slow, and very hard to get a good flat hit on the bug. A flip flop, on the other hand, is light, fast, and flexible.[53] By rapidly accelerating the flip flop, it will bend and squish the bug even from a less than perfect angle of assault.[54] (This is com-

---

[52] Camo's real name is Jared. We nicknamed him Camo because on this trip he wore a camouflage jacket, hat, pants, wallet, underwear, socks, gloves, and a camo tee shirt that said, "Now you can't see me."
[53] It will be understood by the reader that this particular "flip flop" refers to lightweight footwear and not a politician who keeps flexing his position to gain votes on Election Day.
[54] If I had taken physics in New Orleans, I am sure the Professor would have demonstrated the advantage of the flip flop technique while teach-

monly referred to as the Donald Rumsfeld Theory of Bug Squishing. Ol' Rummy was known for using a modern, lighter, faster, more flexible approach to military engagement in Iraq, rather than the "old school" heavy, lumbering, slow-moving assault of previous conflicts. Well, at least it works for bugs.)

While serving in New Orleans one night, we headed into "The Projects" to play games with the hundreds of children who were out enjoying the warm summer evening. Among these children was a beautiful boy about three years old who was wearing a camouflage shirt. I asked for permission to photograph the young boy from "The Projects" with Camo from our group, because of their matching outerwear. I nicknamed the 3-year-old "Baby Camo."

Here is the point.

As I contemplate the inherent value of the boy and the bug, I have to believe that the worth of Baby Camo is infinitely greater than the worth of the bug. If I embrace the perspective of those on Rational Road, then I am forced to conclude that the boy and the bug are ultimately of equal value. They are simply different ends of long sequences of random variations selected for by an impersonal and purposeless process.

Not only must I conclude that they are of equal value, but I must also conclude that their existence has no purpose. No matter how many buttons are on the *smart people's calculators*, my mind simply can not accept the idea that the

---

ing us the equation: Force = Mass x Acceleration. Although the mass of the work boot is significantly greater than the flip flop, this difference is easily offset by the phenomenal acceleration one can apply to the flip flop.

little boy's life from "The Projects" has no more meaning or purpose than that of the bug.

I also think there is a dirty little secret behind the *smart people* on Rational Road who insist there is no Moral Designer. I don't think they really believe it. Otherwise, they wouldn't push so hard for their moral perspective. Why would anyone work so tirelessly toward an end that is without purpose or meaning?

Third, morality will never win over power. If the people on Rational Road are correct, then our morality is something selected for by chance. In order to gain its footing, it had to compete with power. This is the conflict between the "Alpha Wolf" and "Moral" Systems. But morality alone never trumps power.

Let's review Iraq. Harry "We lost the war!" Reid is presently proclaiming it as a moral stand that we should get out of Iraq before losing any more of our people. Purely for the sake of discussion, let's agree with Harry's moral assessment and assertion about the war. Now, does anyone doubt for a moment that we have enough force available to totally dominate that area? So, why are we losing? We are losing because of our morality.

We do not kill innocent people. We ask our soldiers to put their lives at risk. At times they have to make a split second decision that might save the lives of their fellow soldiers. If they make the wrong decision in that split second, they run the risk of being castigated as murderers in the news and being disciplined by the military. Our power is hamstrung by our morality.

We do not wantonly bomb mosques. We respect freedom of religion and want a world where people are allowed to worship according to their conscience. Our power is constrained by our morality.

We do not profile people. Our security system in airports requires us to suspect everyone equally. Forget the statistics on Middle Eastern Muslim men and acts of terror. Grandma gets frisked too! Our power gets diluted by our morality.

Our enemies, who have inferior power, have no such moral restraints. They plot to kill innocent people and rejoice when the numbers of innocent dead skyrocket. They bomb mosques as needed and profile everyone who is not of their sect as worthy of death.

And we are told we are losing the war. Not militarily of course. WE COULD SQUISH 'EM LIKE A BUG WITH A FLIP FLOP! But our morality prevents us from doing so. Their little bit of power wins over our big amount of morality.

Consider partial birth abortion. Does anyone for a minute really believe this can possibly be a moral act under any circumstance?

Let's see! The doctor delivers the healthy baby except for the head. Next he piths it like a frog in a high school biology lab. Then he delivers the remainder of the now dead baby. This is deemed morally acceptable.

However, if he delivers the baby and the head comes out that last four inches, then it is murder if he piths it like a frog. Does anyone truly believe that those last four inches can be calculated to be the difference between morality and murder? Push all the buttons you want, but I don't think so.

This is clearly the question of power winning over morality.

There is a powerful pro-choice movement that has to protect its claim that a woman has the moral right to choose what happens with her body. It is nobody's business to question that decision. If the baby has to give up its life for the benefit of the mother's choice, then so be it.

The heroism of someone sacrificing their life for the benefit of another has been retold in countless stories. We embrace the one making the sacrifice as magnificent in his or her actions and worthy of our respect and admiration. But in partial birth abortion, the sacrifice is demanded of the baby and yet is not something the baby chooses. This is moral? No. This is power. The weak (innocent babies) suffer at the hands of the strong (mothers, the medical community, and pro-choice lobbyists), and morality loses to power.

Although *smart people* may come up with a model which they claim explains how morality supposedly arose in a non-moral "Chemistry and Matter" environment, the model always fails in the real world where morality consistently loses to power.

In fact, the only place where morality wins is when it is backed by power. For example, theft in our culture is only curtailed by locks, security systems, Neighborhood Watch programs, a police force, and a justice system. These power systems form the backbone that enforces our moral system.

Finally, the idea of a Moral Designer, by which morality is Actual, not Virtual, is the only way we can live consistently. Even people who travel Rational Road constantly feel compelled to demand moral standards, though they know there is no ultimate purpose or meaning to give substance to their standards. Christopher Hitchens places a moral judgment upon Jerry Falwell. Chris claims Jerry's life's work was "evil" and thus in conflict with his perception of morality. Michael Newdow, famed atheist who wants to remove "under God" from the Pledge, stakes his claim on his supposed "right" to zero influence from religion in the public sector. The very idea of a "right" presupposes the existence of a moral standard. "Chemistry and Matter" are not moral, nor do they have rights in and of themselves.

If the Rational Road traveler is going to be consistent, he must always remind himself that the *Moral Outrage!* which is driving his passions has no meaning. He is compelled to live for, fight for, and die for a contrived moral standard that is ultimately meaningless beyond his insignificant personal opinion. What an unpleasant condition it must be to be driven by *moral passions which cry out for justice,* while at the same time have one's mind constantly whispering that one's ideas of justice, right and wrong, and good and evil are ultimately ideas that don't actually exist. No wonder Ol' Chris was so angry with the deceased Reverend. Who could ever be happy with that kind of inner conflict? How does one find joy when our very essence (a moral being) must be denied as having any significance whatsoever?

In the great animated film *The Lion King*, Rafiki, the Baboon Prophet, sought out Simba and confronted him with his moral responsibility to protect the pride's land. Simba had been running from his own past and identity. He was pretending that everything was okay about his life as he hung out with Timone and Pumba doing absolutely nothing of purpose (Hakuna Mutata!). But deep inside things were not right. Rafiki promised to show Simba his father, Mufasa. He brought Simba to a pond and told him to look into it. When Simba protested that he was merely seeing his own reflection, Rafiki told him to look closer. Eventually, Mufasa appeared to Simba in his own reflection and they spoke with one another. Mufasa challenged Simba to remember who he was. At that point, Simba ran off to claim his identity and his place as The Lion King. It was only in embracing the reality that he was the son of his father, that Simba found the joy, courage, and will to live with purpose. So it is with humans. It is only in embracing the reality that we bear the imprint of a Moral Designer that we find the joy, the courage, and the will to live our lives with the moral purpose for which they were intended.

Shoot! If an animated baboon can figure this out, we ought to be able get it right too!

****Newsflash!****

Paris is getting out of jail for real this time! Having served her time and paid her debt to society, she is soon to be released.

Word on the street is that she is shopping around for her exit interview to be over a million dollars. That means she will make about $50,000 per day for her time spent in the slammer.

What a great way to pay your debt to society. I wish I could pay my debts and make a huge profit, too!

The obligatory *Moral Outrage!* being expressed by the newscaster is that it is wrong for her to shop her story around. She should not be making money off of this experience. She already has enough. But if she gave her profits to charity, then they would consider this morally acceptable. (Clearly this newscaster's network wasn't the highest bidder on Paris's story.)

Oh, by the way! In case you were wondering, in order to maintain their neutrality, the News Divisions of the networks never pay for an interview. All interviews are purchased out of the Sports budget. That way they can maintain their policies that the News Division NEVER PAYS FOR AN INTERVIEW.

I feel so good knowing these morally upright individuals, who pontificate about Paris' immoral actions while making their living off of reporting about her, would never do anything even a little bit morally askew. They are SO TRUSTWORTHY!

Chapter 39

Moonwalker Conspiracies

In my lifetime there have been two notable Moonwalkers.

The first was some astronaut[55] who landed on the moon and said those famous words as he was taking Man's first steps on the lunar surface, "Wow! Talk about a Road Less Traveled!"

The second notable Moonwalker was a Pop Idol from my childhood. He was singing on the Ed Sullivan show at the age of five. He was already rich and famous by then. I was about eight and can remember my Dad watching him and saying to me, "Hey, if you were rich and famous I could be

---

[55] I would probably have his name forever emblazoned in my mind if I actually watched this thing on television when it was happening, but being a typical American kid at the time, I had more important things to do than watch this incredible historic event unfold before my eyes. I was playing outside. This is no longer considered typical behavior. Now kids need to be told by famous athletes on TV that they are to go outside and play.

retired already." His name is Michael Jackson. Actually he wasn't Moonwalking at the age of five yet. At this point, he was just getting started on hiring lawyers, personal investors, chauffeurs, agents, bodyguards, and plastic surgeons. He began buying houses, limos, clothes, and influence. He was also laying out his plans for a place called Neverland. It was his dream to create a place for kids where he could steal the beauty of their youth to replace the one he never had.

While both of these guys are known for Moonwalking, they share another common experience. Both of them had people respond to their situations with total disbelief. Whatever the motives behind their assertions, there are those who insist until this day that the government perpetrated a fraud upon the American people about landing on the moon. Others believe that Jacko is completely innocent and has nothing but love for the children.

Likewise, there will always be those who insist that there is no Moral Designer behind their deep desire for what is just, good, and right in the world. They ignore the reality of living completely illogical and inconsistent lives. Traveling gleefully down Rational Road, they simply will not accept the idea of a Moral Designer any more than some people will accept that men ever walked on the moon. But they will insist that their *Moral Outrage!* must be heard, because this is morally right. (Do I need to repeat yet again that the idea of a moral "right and wrong" makes no sense in a world that is solely the result of random chemical reactions?)

Now, we should fault no one for insisting that moral standards must prevail. That is the point of this whole discussion. *The essence of our humanity is found in our morality.* It is what sets us apart as unique in the animal kingdom.

We all believe that there is a moral "right and wrong," and we all live by some code of conduct that we believe reflects that moral standard. We cannot escape living this way, for to do so would be to deny our very being.

This brings us back to the beginning of our quest.

When we watch the constant banter of *Moral Outrage!* expressed by the politicians, party leaders, pundits, talking heads, bloggers, commentators, talk radio people, union negotiators, jihadis, religious leaders, environmentalists, movie stars, atheists, televangelists, philosophers, editorialists, news reporters, war protesters, animal rights activists, women's rights activists, pro-life activists, gay rights activists, traditional marriage activists, minority rights activists, open border activists, closed border activists, etc., etc., etc., how do we know any of them are correct in the moral stand they propound?

I have run this through my slide rule a number of times and the only way we can know for sure who is right is *if a Moral Designer has told us what the good and perfect Moral Standard truly is.*

Then it becomes our moral responsibility to understand the Perfect Standard and seek to get as close to it as possible in our laws and customs.

If you will indulge my repeating one last time, without a Standard revealed by a Moral Designer, our *Moral Outrage!* is little more than our personal opinions. Although we may feel good about forcing our opinions on someone else (who doesn't feel good when they get their way?), we can't honestly say we are holding to a moral position. At best, we can only say that we have the power to make things go the way we *think* is right. But simply because we have the power, and simply because we think our standards are morally just, that does not make it so.

And this pushes us to the ultimate philosophical question of morality:

HAS A MORAL DESIGNER MADE KNOWN TO US THE CORRECT MORAL STANDARD?

That philosophical question immediately leads to two practical ones.

If no Moral Designer has revealed a Moral Standard, to whom or to what do we turn to resolve our *Moral Outrage!*?

If a Moral Designer *has* revealed a Moral Standard, shouldn't we have some way of measuring how the politicians and opinion makers square up with The Standard?

If we are going to put our trust in all these movers and shakers to solve our problems and bring the solutions to the *Moral Outrage!* that pervades every issue in our country, shouldn't we have some way of measuring how they match up with The Standard defined by The Moral Designer?

Oh, wait! That would require us to be able to talk about The Standard. That might make someone who wants to deny there is a Standard feel uncomfortable. So, as a nation, we will ignore The Standard and pretend that our personal opinions, posturing, and *Moral Outrage!* are adequate to keep us functioning as a morally upright people.

The only problem with this little national self-deception is that Morality must instruct Power in order for Power to not degrade into corruption. Since we have insisted that The Standard cannot be a part of our National Discussion, we have placed ourselves on an inevitable path towards decline, because no one has the courage or the legal standing to invoke The Standard as a means of keeping us upright.

That means that every election year, we are being asked to invest incredible power into the hands of people who will not have to operate according to The Standard. Even this slide rule toting, likes to laugh at funny movies, partial credit, graded on a curve, low-tech, Lincoln Log builder, and traveler on Designer Drive knows that there are precious few people in the world who can handle that kind of power without accountability. Most humans quickly become adept at manipulating their power to apply a "double standard" instead of The Standard. They will proclaim their *Moral Outrage!* at the behavior of those with differing opinions, yet wink at the identical behavior in those who share their opinions.[56] They can get away with this because we have long since abandoned any standard (Constitutional or Moral) to which they are all held accountable. Once they have their power, they will use their power to maintain their power. They will do this with a straight face[57] knowing there are no consequences, because with their power they control the consequences.

And so we are left with a moral dilemma with no chance to make "dilemmanade."

It's more like we are all asked to drink da' Kool-Aid.

---

[56] Think Scooter Libby vs. Sandy "Burglar"
[57] Well, they don't always have a straight face. While expressing *Moral Outrage!* toward those of an opposing opinion, they speak with contortion and anger so you know they are sincere. The straight face is only used when they are overlooking the moral inadequacies of their own compadres.

We are expected to believe that the *Moral Outrage!* expressed by the politicians, pundits, activists, commentators, entertainers, editorialists, and talking heads means these people are qualified to make moral decrees.

But *Moral Outrage!* does not equal moral certainty.

AND WITHOUT A STANDARD, MORAL CERTAINTY CANNOT BE ASCERTAINED.

So, unless we are bold enough to embrace The Standard, we are resigning ourselves as a nation to being ruled by nothing more than the opinions of those who happen to be able to gather the most power to enforce their opinions upon us.

And by George, I just haven't yet seen the person to whom I want to entrust all that power so they can enforce their opinions upon me.

While flailing about in this emotional and intellectual quagmire, I have been confronted with another unbelievable development. One more financial institution has gotten on the Monkey Bandwagon in anticipation of this project being published. Rather than using a cute chimp, baby orangutan, or little capuchin, they have gone with the "Bigger is better" approach. They utilize a talking gorilla giving financial advice. When totally ignored, the gorilla quips, "But what would I know? I'm just the 800-pound gorilla in the room."

And everybody is ignoring the 800-pound gorilla in our nation's living room today.

Our national discourse overflows with *Moral Outrage!*, but our practice wallows in things far less upright.

We claim we are striving to be a moral society.

But we teach our children that their very existence is owed to a totally non-moral, random process.

We claim we are striving to be a moral society.

But we insist our public discourse must be devoid of any reference to or influence from a Moral Designer.

We claim we are striving to be a moral society.

But we insist that our historical standard for morality must be eradicated from our discourse, lest it actually bear influence upon someone's thinking.

We claim we are striving to be a moral society.

But we have morphed into moral and intellectual Neanderthals, because we have forgotten how to discern between establishing laws based upon true morality and forcing people to embrace a world view against their will.

We claim we are striving to be a moral society.

But the 800-Pound Gorilla in our nation's living room is that we have abandoned the very foundations and Standard of morality that made our nation great. And we have arrogantly come to believe that we can independently be the ultimate discerners of morality in our universe.

Perhaps the real truth is that our arrogance has so deluded us we are as morally naked as the Emperor with no clothes.

Is anyone else feeling a draft around here?

Chapter 40

****Final Newsflash!****

This project is finally done. But not to worry!

Our *Moral Outrage!* is continually being stoked by many things that entertain us and make us feel morally superior.

The following stories have all been in the news in the past 48 hours.

Paris Hilton is releasing a new DVD with never-before-seen images of herself!

Lindsey Lohan was arrested for a DUI, suspended license, and cocaine possession.

Britney Spears had a major meltdown at a photo shoot with people alluding to the possibility of substance abuse during the shoot.

An NBA Referee is being investigated for shaving points for the Mob.

Michael Vick, NFL quarterback, is in deep doggy doo-doo over dog fighting allegations.

Barry Bonds is two home runs away from Hank Aaron's home run record with questions surrounding his possible use of steroids.

These issues provide the talking head moralists with endless material upon which to pontificate concerning the "right or wrong" of each circumstance. I will leave them to themselves to pronounce their righteous judgments. Apparently, they can do this because they are the self-appointed purveyors of all moral truth.

And I will leave you, the reader, with three questions to ponder:

If our existence is due solely to random processes of nonmoral "Chemistry and Matter,"
    And if there is no Moral Designer,
    And if there is no Moral Standard,
Who is to say and on what basis do they say it, that any behavior is wrong or immoral?

If morality is "Actual," *where did it come from?*

If morality is "Virtual," *why do we experience such persistent intense Moral Outrage!?*

The Evolution of Calculators as Pick-Up Tools
Push-Button Calculator ~ The "Upside-Down Numbers" Method

The Male Mind After a Gazillion Years of Evolution. Disappointing.

Chapter 41

Genetic Markers

Pluto and Partial Credit

\# *Newsweek*, September 4, 2006, p. 50
* Ibid., p. 47
\+ Ibid., p. 50

How to Spot Smart People

\# *Time,* October 9, 2006, p. 47
* *Sideways*, 2004, 20th Century Fox Film Corporation
\+ *Time*, October 9, 2006, p.50

Symbiosis

\# *Newsweek*, September 4, 2006, p. 50
* *Time*, October 9, 2006, p. 48

Opposable Thumbs

\# *Time*, October 9, 2006, p. 46

The Pluto Problem

Time, October 9, 2006

All Politics Are Local

PEZ and PEZ Candy Dispenser are registered trademarks of PEZ Candy Company, Inc.

The Continental Divide

PEZ and PEZ Candy Dispenser are registered trademarks of PEZ Candy Company, Inc.

Smart Wars

Newsweek, March 19, 2007, p. 57
* Ibid., p. 58
+ Ibid., p. 56

Enigmas of Virtual Morality

Fresh Air, Minnesota Public Radio, March 27, 2007
* Ibid.
+ Fox News Television, May 24, 2007

Chapter 42
Vestigial Organ

[The following was a chapter entitled **Kenomorality** in the original text of this work. The purpose of this chapter was to introduce the term "kenomorality" into the English lexicon. But after coining (so I thought) the synonymous term "Virtual Morality," it became clear this chapter was no longer needed and as such, it became a proverbial "vestigial organ."]

Moral Outrage! is the foundation of all of our public debate. Every *smart person proponent* of any particular viewpoint on any subject is basing their opinion on their personal perspective of morality. If you disagree with them, you, by their definition, are embracing immorality. And because they totally believe they are right, and because they believe so strongly in their sense of right and wrong, they passionately hold to their position. Watch *The O'Reilly Factor* for a clear example of this dynamic.

But even The Cavemen from the *Geico* commercials could

connect the dots that say this is all ridiculous. If we are only "Chemistry and Matter," and "Chemistry and Matter" are amoral in and of themselves, and if "Chemistry and Matter" being amoral cannot create morality, then all of these discussions about moral high ground, *Moral Outrage!*, and accusations of immorality (whether in one's private or public life) really amount to nothing.

Morality is not a reality!

So, while we are all trying to live on the "moral high ground," why don't we start with honesty? (That is a commonly held "moral position," isn't it?) Rather than claiming that our opinions are rooted in "morality," let's fess up and admit that our opinions are at best rooted in "kenomorality."

Smart people linguistic types say "kenomorality" is from the Greek "kenosis" which means "to empty." Therefore, it literally means "empty morality." It is morality that claims "right and wrong" even though it knows that "Chemistry and Matter" are not actually moral. The foundation of moral claims is in fact non-existent. It is empty.

Not being very smart myself, I prefer to think of "kenomorality" as derived from Ken of "Barbie and Ken" fame. His "Chemistry and Matter" evolved from the plastics industry and has no moral significance. He just seems like something that "should" be there (Barbie "should" have a boyfriend), but he doesn't really matter (Barbie can have all of her toys and her exciting life totally without Ken). Morals seem like something that "should" be there (we "should" protect the weak and vulnerable), but ultimately, if we are only "Chemistry and Matter", morals do not really matter (so what if the weak and vulnerable are preyed upon by the strong?).

So, let's have our moral discussions. Let's hold our opinions. Let's make our accusations. Let's just remember what the Biker Philosopher said. Iddohmadduh! It's like sports. We can argue whatever position we want about who is better, but when all is said and done, it amounts to nothing.

"Kenomorality." Emotionally, it acknowledges our moral compass while intellectually allowing us to remain honest.

And once we embrace "kenomorality," physics class can be a whole lot more fun. Once we embrace the fact that morality ultimately doesn't matter, the professor can start shooting darts at real monkeys. Imagine how loud they'll squeal when that dart catches them in the crotch!

About the Author

Raised in the middle class suburbs of Chicago and presently living in the middle class suburbs of a town of 362 people somewhere in "flyover country", educated in a liberal world view at a Big Ten state university and a conservative world view at a private graduate school; having served the wealthy in America and the poorest of the poor in a third world country; Gary has learned that life can be a wonderful array of contrasts. But for all the satisfaction that life's experiences may bring, he has found there is no greater joy than the simple and profound pleasure of having his three children around the dinner table to share the Holidays with him and his wife, Laurie.

Printed in the United States
130524LV00002B/3/P